SURVEY
ORGANIC
CHEMISTRY
LABORATORY MANUAL

SIXTH EDITION

ARDESHIR AZADNIA

Kendall Hunt
publishing company

Kendall Hunt
p u b l i s h i n g   c o m p a n y

www.kendallhunt.com
Send all inquiries to:
4050 Westmark Drive
Dubuque, IA 52004-1840

Copyright © 1999, 2002, 2005, 2009, 2012, 2016 by Ardeshir Azadnia

ISBN 978-1-5249-0050-2

Printed in the United States of America

Contents

Introduction

Many introductory organic laboratory students feel that the laboratory does not serve any useful purpose for them. They regard the techniques as too obscure and the material too difficult. For this introductory organic laboratory, we have done our best to ensure that the concepts to which you are exposed, and the ideas that we wish to get across, are relevant to introductory non-major organic lab students. Many of the techniques you will be exposed to utilize the same basic principles as techniques used in biochemical and clinical laboratories. The purpose of this laboratory is twofold: to give you an introduction to basic organic chemistry laboratory techniques and second, and perhaps more important, to give you some idea of how chemistry (and science in general) works in a laboratory setting. Each experiment is designed with this concept in mind. Each new technique is first demonstrated in its most basic form, and then it is applied to a practical laboratory use.

You will also learn to have respect for chemicals and how to handle some hazardous materials. In addition, in this course you should learn to use prudent laboratory practices, including proper disposal of hazardous chemicals. This is extremely important for keeping our environment clean.

We hope you will find CEM 143 both challenging and stimulating. We hope that this hands-on experience in organic chemistry shall benefit you in your future endeavors.

Check-in Procedures

Record your locker number in your notebook. Carefully check all of your apparatus, washing anything that is not clean. A distilled water rinse will prevent watermarks. Replace any chipped, cracked, broken, or missing apparatus before starting any experiments. See your instructor for details on how to obtain replacement items. Give your instructor your completed inventory sheet.

Before the first meeting of your laboratory section you must go to the chemistry department safety website. Read the safety regulations carefully and provide all of the requested information; you will not be allowed to check in to your locker until you have completed the online safety sheet. Your locker may be reassigned if the check-in procedure is not followed. Be sure that all paperwork is completed and returned the first day.

Safety Regulations

In order to avoid personal injuries and injuries to fellow students while performing experiments in your Chemistry Laboratory Courses, it is required that you read and understand the following regulations before performing any experiments. The department reserves the right to exclude any person from the laboratory who endangers him/herself or others.

A. Personal Protection

1. Approved safety goggles (not sunglasses) must be worn at **all** times when in the laboratory. *Soft contact lenses shall not be worn in the laboratory* under any circumstances, even under goggles. Hard contact lenses are conditionally acceptable. However, you must first check with your instructor.

2. If you get a chemical in your eye, immediate and extensive washing for 15 to 20 minutes with water **only** is absolutely essential to minimize damage. Use an eye wash bottle, a hose, an eye fountain, or an eyecup at once. If you spill any chemical on yourself, immediately wash with large amounts of water; then notify your instructor.

3. The wearing of rubber gloves and aprons is strongly advised when working with toxic and/or corrosive substances. However, *gloves must never be a substitute for neatness and careful technique.* Do not use organic solvents to remove organic compounds from the skin: they will only spread the damage over a wider area. Solvents also tend to penetrate skin, carrying other chemicals along. Soap and water are more effective.

4. Do not apply ointments to chemical or thermal burns. Use only cold water.

5. Do not taste anything in the laboratory. (This applies to food as well as chemicals. Do not use the laboratory as an eating place and do not eat or drink from laboratory glassware.) Do not use mouth suction in filling pipettes with chemical reagents. (Use a suction bulb.)

6. To minimize hazard, confine long hair securely when in the laboratory. Also, a laboratory apron is essential when you are wearing easily combustible clothing, especially synthetics. Such an apron affords desirable protection on all occasions. Properly closed shoes or sneakers (i.e., no sandals) and long pants must be worn in labs at all times.

7. Exercise great care in noting the odor of fumes and whenever possible avoid breathing fumes of any kind. See also C-6.

8. No smoking in labs. MSU is a smoke-free campus.

9. Report all cuts, burns, and other laboratory injuries to your instructor at once. You are advised to obtain medical attention for cuts, burns, inhalation of fumes, or any other laboratory incurred accident. If needed, your laboratory instructor will arrange for transportation to Olin Health Center. An accident report must be completed by your laboratory instructor.

10. No earphones or earbuds of any kind shall be worn in laboratories.

B. Property Protection

1. In case of a fire, call the instructor at once. If you are near an extinguisher, bring the extinguisher to the fire, but let the instructor use it.

2. Know the location of all safety equipment: fire extinguisher, safety showers, fire blankets, **eye–wash stations,** and exits.

3. Treat all liquids as extremely flammable unless you know them to be otherwise.

4. Report all spills promptly to your laboratory instructor. You will be advised of the proper cleanup procedure.

5. Disposal of hazardous waste: dispose of all chemicals, liquid and solid, into the proper waste containers. Ask your instructor how to dispose of waste chemicals you are unsure about.

6. Place broken glass in the appropriate container. Do not put broken glass in the wastepaper cans.

C. Laboratory Technique

1. Read and study the experiment before coming into the lab. This will allow you to plan ahead so that you can make best use of your time. The more you rush at the end of a lab, the greater your chance of having an accident.

2. Perform no unauthorized experiments. Do not remove any chemicals or equipment from the laboratories. You alone will bear the consequences of "unauthorized experimentation."

3. Never work in any laboratory alone!

4. Do not force glass tubing into rubber stoppers. (Protect your hands with a towel when inserting tubing into stoppers, and use a lubricant with a twisting motion.)

5. When working with electrical equipment observe caution in handling loose wires and make sure that all equipment is electrically grounded before touching it. Clean up all puddles immediately.

6. Use hood facilities. Odors and gases from chemicals and chemical reactions are usually unpleasant and in many cases toxic.

7. View reactions from the side, keeping your safety glasses between you and the reactants. Do not look into the open mouth of a test tube or reaction flask. Point the open end of the tube away from you and other laboratory workers.

8. Be a good housekeeper. Order and neatness will minimize accidents.

9. Laboratory safety is the personal responsibility of each and every individual in the laboratory. Report unsafe practices.

Treat all chemicals as corrosive and toxic and all chemical reactions as hazardous unless you know them to be otherwise.

Rubber Gloves

The greatest hazards associated with chemicals generally concern swallowing or inhaling vapors, fumes, or mists. Many people consider only corrosive materials such as lye or sulfuric acid dangerous to skin. This is a false assumption. Skin resistance to chemicals varies. In some cases skin resistance is very good, but for others, especially lipid soluble materials, absorption through the skin can produce dangerous levels in the body. Some common chemicals, which are readily absorbed through the skin in toxic amounts, include:

Aniline	Cyanides	Mercury
Benzene	1,2-Dibromoethane	Nicotine
Bis(chloromethyl) ether	N,N-Dimethylaniline	Nitrotoluene
Bromoform	Dimethylformamide	Phenol
Carbon tetrachloride	Hydrazine	Tetraethyl lead

As time passes, each chemist becomes aware that things in the laboratory are not always simple or safe. Rubber gloves are of this class of items. The following generalizations about each glove material can be made:

Nitrile: a copolymer of butadiene and acrylonitrile. Noted for its resistance to puncture, abrasion, and most chemicals, particularly petroleum solvents, oils, acids, caustics, alcohols.

Neoprene: a polymer of 2-chloro-1,3-butadiene. The standard for glove boxes, recommended for oils, greases, gasoline, DMF. **NOT** for use with aromatics and chlorinated hydrocarbons, and strong oxidizers.

Natural Rubber: excellent for use with alcohols and caustics. Good for DMSO and aniline, also most ketones. These are the commonly seen thin, tan colored gloves. They are rapidly destroyed by thionyl chloride and chlorosulfonic acid.

Butyl Rubber: most impermeable to gases and water vapor. Best for aldehydes and ketones, caustics, and amines. Generally good all-around protection except for aromatics, chlorinated hydrocarbons, and petroleum solvents.

PVC: polyvinyl chloride supported gloves are generally best for inorganic and organic acids, caustics.

Polyethylene: these are always disposable; excellent for acids, caustics, aldehydes, and ketones.

PVA: polyvinyl alcohol supported gloves are **the** best for aromatic and chlorinated hydrocarbons, ketones, THF, but cannot be used with aqueous systems as they dissolve in water. Also not good for DMSO, DMF, and pyridine.

Most of these gloves may be purchased at any grocery or drug store.

Safety Data Sheet

The chemical industries are mandated to provide the end user with sufficient information about the hazards associated with a particular chemical. This information is called a Safety Data Sheet (**SDS**). SDS informs one of chemicals' physical properties, flammability, corrosiveness, combustibility, and other hazards such as absorption through skin, etc. It is extremely important for researchers to read the SDS information for the chemicals that they are about to use prior to doing the reaction in order to take the appropriate precautions. SDS for all chemicals that are being used in the chemistry department can be found by doing the following:

Go to HYPERLINK "http://www.chemistry.msu.edu", click on safety, click on SDS, and type the chemical name.

NFPA CHEMICAL HAZARD LABELS

| Identification of Flammability | | Identification of Reactivity | | (Stability) | |
| Color Code: Blue | | Color Code: Red | | Color Code: Yellow | |
Type of Possible Injury	Signal	Susceptibility of Materials to Burning	Signal	Susceptibility to Release of Energy	Signal
4	Materials, which on very short exposure, could cause death or major residual injury even though prompt medical treatment was given.	4	Materials which will rapidly or completely vaporize at atmospheric pressure and normal ambient temperature, or which are readily dispersed in air and which will burn readily.	4	Materials which in themselves are readily capable of detonation or of explosive decomposition or reaction at normal temperatures and pressures.
3	Materials, which on short exposure, could cause serious temporary or residual injury even though prompt medical treatment was given.	3	Liquids and solids that can be ignited under almost all ambient temperature conditions.	3	Materials, which in themselves, are capable of detonation or explosive reaction but require a strong initiating source or which must be heated under confinement before initiation or which react explosively with water.
2	Materials, which on intense or continued exposure, could cause temporary incapacitation or possible residual injury unless prompt medical treatment is given.	2	Materials that must be moderately heated or exposed to relatively high ambient temperatures before ignition can occur.	2	Materials, which in themselves are, normally unstable and readily undergo violent chemical change but do not detonate. Also materials which may react violently with water or which may form potentially explosive mixtures with water.
1	Materials, that on exposure, would cause irritation but only minor residual injury even if no treatment is given.	1	Materials that must be preheated before ignition can occur.	1	Materials, which in themselves are normally stable, but which can become unstable at elevated temperatures and pressures or which may react with water with some release of energy, but not violently.
0	Materials, which on exposure under fire conditions, would offer no hazard beyond that of ordinary combustible material.	0	Materials that will not burn.	0	Materials, which in themselves are normally stable, even under fire exposure conditions, and which are not reactive with water.

Special Notice Key (Color Code: WHITE)

W—water reactive

Oxy—oxidizing agent

Common SDS Terms

Acute Effect—An adverse effect with severe symptoms occurring very quickly, as a result of a single excessive overexposure to a substance.

Acute Toxicity—The adverse effects resulting from a single excessive overexposure to a substance. Usually a figure denoting relative toxicity.

Asphyxiant—A vapor or gas that can cause unconsciousness or death by suffocation. Most are associated with a lack of sufficient oxygen to promote life.

Asphyxiation—Is the starvation of the body for oxygen.

Boiling Point—A temperature at which a liquid turns to vapor. This term is usually associated with the temperature at sea level pressure when a flammable liquid gives off sufficient vapors to promote combustion.

"C" or Ceiling—In terms of exposure concentrations, this is the number that should never be exceeded even for a short period, for a substance.

Carcinogen—A substance or agent capable of producing cancer in mammals.

cc (cubic centimeter)—A volume measurement usually associated with small quantities of a liquid. One quart has 946 cubic centimeters.

Chronic Effect—An adverse effect with symptoms that develop or recur very slowly, or over long periods of time.

Chronic Toxicity—The adverse effects resulting from prolonged or repeat exposures to a substance, usually used as an indicator of relative toxicity for exposures over great lengths of time.

Combustible—A term used to classify liquids, gases, or solids that will burn readily. This term is often associated with "flash point," which is a temperature at which a given material will generate sufficient vapors to promote combustion.

Concentration—A figure used to define relative quantity of a particular material.

Corrosive—A material with the characteristics of causing irreversible harm to human skin, or steel by contact. Many acids are classified as corrosive.

Decomposition—The breakdown of materials or substances into other substances or parts of compounds. Usually associated with heat or chemical reactions.

Dermal—Used on or applied to the skin.

Dermal Toxicity—The adverse effects resulting from exposure of a material to the skin. Usually associated with lab animal tests.

Evaporation Rate—The rate at which a liquid material is known to evaporate, usually associated with flammable materials. The faster a material will evaporate, the sooner it will become concentrated in air, creating either an explosive/combustible mixture or toxic concentration, or both.

Flash Point—The temperature at which a liquid will generate sufficient vapors to promote combustion. Generally, the lower the flash point, the greater the danger of combustion.

Flammable—Any liquid that has a flash point of 100°F or below. Also, any solid, which can sustain fire and ignite readily.

General Exhaust— A term used to define a system for exhausting or ventilating air from a general work area. Not as site specific as localized exhaust.

"g" (gram)—A unit of weight. One ounce equals about 28.4 grams.

Hazardous Chemical—Any chemical which is either a physical or health hazard or both.

IDLH (Immediately Dangerous to Life and Health)—These are concentrations which, if breathed continuously for at least one-half hour in the air, could cause irreparable damage to most adults.

Ignitable—A term used to define any liquid, gas, or solid, which has the ability to be "ignited" which means having a flash point of 140°F, or less.

Incompatible—Materials, which could cause dangerous reactions from direct contact with one another.

Ingestion—Taking in of a substance through the mouth.

Inhalation—The breathing in of a substance in the form of a gas, liquid, vapor, dust, mist, or fume.

Inhibitor—A chemical added to another substance to prevent an unwanted change from occurring.

Irritant—A chemical, which causes a reversible inflammatory effect on the site of contact, however, is not considered a corrosive. Normally irritants affect the eyes, skin, nose, mouth, and respiratory system.

LC (Lethal Concentration)—In lab animal tests, this is the concentration of a substance which is sufficient to kill the tested animal.

LC_{50} (Lethal Concentration$_{50}$)—In lab animal tests, this is the concentration of a substance required to kill 50% of the group of animals tested.

LD (Lethal Dose)—The concentration of a substance required to kill the lab animal used to the test with a specific material.

LD_{50} (Lethal Dose$_{50}$)—The single dose concentration of a substance required to kill 50% of the lab animals tested.

LEL (Lower Explosive Limit, older term for LFL used in literature)—The lowest concentration, or percentage in air, of a vapor or gas, that will produce a flash fire when an ignition source is introduced.

LFL (Lower Flammability Limit)—It is the lowest composition of fuel in air that will burn with a flame in the absence of a continuous external source of heat.

Local Exhaust—The system for ventilating or exhausting air from a specific area such as in welding operations. More localized than general exhaust.

Melting Point—The temperature at which a solid changes to a liquid.

mg (Milligram)—A unit of measurement of weight. There are 1000 mg in one gram of a substance.

mg/m³ (Milligrams Per Cubic Meter)—A unit of measurement usually associated with concentrations of dusts, gases, or mists in air.

mppcf (Million Particles Per Cubic Foot)—A unit of measure usually used to describe airborne particles of a substance suspended in air.

Mutagen—A substance or agent capable of altering the genetic material in a living cell. Normally associated with carcinogens.

NFPA (National Fire Protection Association)—An organization which promotes fire protection/prevention and establishes safeguards against loss of property and/or life by fire.

The NFPA has established a series of codes identifying hazardous materials by symbol and number for firefighting purposes. These codes also classify materials in their order of flammability, with 0 being not burnable up to 4, which means will burn spontaneously at room temperature. A more detailed description of NFPA Symbols is attached.

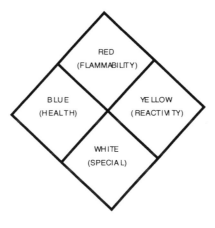

NFPA SYMBOL

Olfactory—Relating to the sense of smell.

Oral—Used in or taken through the mouth into the body.

Oral Toxicity—A term used to denote the degree at which a substance will cause adverse health effects when taken through the mouth. Normally associated with lab animal tests.

Oxidizer—A substance which yields oxygen readily to stimulate the combustion of an organic material.

Oxidizing Agent—A chemical or substance which brings on oxidation reactions (i.e., by providing the oxygen to promote oxidation).

PEL (Permissible Exposure Limit)—An exposure concentration established by the Occupational Safety & Health Community, which indicates the maximum concentration for which no adverse effects will follow.

PPB (Parts Per Billion)—As above, only expressed as number of parts per billion parts of air.

PPM (Parts Per Million)—A unit of measurement for the concentration of a gas or vapor in air. Usually expressed as number of parts per million parts of air.

Pyrophoric—A chemical that will ignite spontaneously in air at a temperature of 130°F (54.4°C) or below.

Reactivity—The term which describes the tendency of a substance to undergo a chemical change with the release of energy, often as heat.

Reducing Agent—In an oxidation reaction, this is the material that gets oxidized.

Respiratory System—The breathing system, including the lungs and air passages, plus their associated nervous and circulatory components.

Sensitizer—A substance, which on first exposure causes little or no reaction, however, with repeated exposure would induce a marked response not necessarily limited to the exposure site. Usually associated with skin sensitization.

Specific Gravity—The weight of a material compared to the weight of an equal volume of water. Usually expresses a materials density. A material with a specific gravity of greater than 1.0 will sink to the bottom in water, whereas a material with a specific gravity of less than 1.0 will float on top of water.

STEL (Short Term Exposure Limit)—The maximum allowable concentration of a substance that one can be exposed to for less than 15 minutes and not produce adverse health effects.

Teratogen—A substance or agent, usually associated with cancer, that when exposed to a pregnant female will cause malformation of the fetus. Usually associated with lab animal tests.

TLV (Threshold Limit Value)—A term used by the Occupation Safety & Health Community to describe the airborne concentration of a material to which nearly all persons can be exposed to day in and day out, and not develop adverse health effects.

Toxicity—The sum of adverse effects of exposure to materials, generally by mouth, skin, or respiratory tract.

TWA (Time Weighted Average)—The airborne concentration of a material to which a person can be exposed over an 8-hour work day (an average).

UEL (Upper Explosive Limit, older term for UFL used in literature)—The highest concentration of a gas or vapor in air that will sustain or support combustion when an ignition source is present.

UFL (Upper Flammability Limit)—The composition above, which the fuel will not burn in air.

Vapor Density—A term used to define the weight of a vapor or gas as compared to the weight of an equal volume of air. Materials lighter than air have a vapor density of less than 1.0, whereas materials heavier than air have a vapor density greater than 1.0.

Vapor Pressure—A number used to describe the pressure that a saturated vapor will exert on top of its own liquid in a closed container. Usually, the higher the vapor pressure, the lower the boiling point, and therefore the more dangerous the material can be, if flammable.

Waste Disposal

A chemical laboratory or manufacturing plant is the potential source of numerous byproducts and hazardous chemical wastes. The impact of these materials on the environment can be greatly reduced or eliminated altogether by following appropriate procedures.

Not all wastes are treated in the same manner. Wherever possible, laboratory wastes will be processed to recover useful materials. Materials that cannot economically be recycled are generally disposed of by incineration, possibly with special scrubbing of flue gases and by burial in approved landfills. Some wastes are readily biodegradable, and small amounts may be flushed down the drain.

The ecologically safe disposal of chemical wastes and by-products requires the cooperation of **everyone** in the lab. Place the waste materials in the liquid hazardous waste container or the solid hazardous waste container, as appropriate; unless you are specifically told that it can go down the drain or into the ordinary trash. Remember also that hazardous reactions may occur when different types of waste are mixed. Follow the directions of this manual and your instructor carefully; read all labels twice. Know what you are discarding.

Broken glassware and stoneware belongs in the red BROKEN GLASS buckets. Do not put sharp objects into the wastepaper cans where they might injure an unsuspecting custodian.

Housekeeping

Each student should sponge off his/her work area at the end of each lab period. All community equipment (clamps, rods, hoses, ice baths, steam bath kits, etc.) are to be returned to the proper compartments at the end of each lab period. In particular, each bench space should have a steam bath with both hoses and a **complete set of rings**. Hoarding community apparatus is grounds for lowering your grade.

Community areas such as balances are to be kept clean. If you spill chemicals, report it to your TA immediately. Do not try to clean it yourself. The lab instructors are encouraged to assign cleanup duties if needed. If you feel that the lab is not clean when you enter, tell your instructor promptly. Appropriate penalties will be levied against the offenders. Remember that the next section will evaluate your housekeeping also!

On Keeping a Laboratory Notebook and Write Ups

No post-lab reports will be required. However, you should have carefully studied the assigned sections of your text before coming to lab. This will permit you to work much more efficiently in lab. **The laboratory manual is not to be used in lab**—this encourages better preparation.

An important laboratory skill is the making and recording of accurate observations. A well-kept notebook is an essential part of any investigation. The laboratory notebook must be a permanently bound, prenumbered book, with perforated pages for the carbon copies. The **carbon copies** are to be given to your instructor at the **end of each lab period**. Each page should be signed and dated. **Part of your pre-lab preparation should consist of recording in your notebook exactly what you plan to do during your next lab period**. During the lab period, you should record all observations you make and report any changes made in the procedure. Someone should be able to get precisely the same results that you obtained by following only your notebook. Your notebook should be a complete description of what happened—or did not happen—in your experiments.

1. Place 5.0 gm of acetic acid in a 100 mL round bottom flask.

2. Add 50 mL of ethanol.

3. Add 5 drops of conc. H_2SO_4.

4. Attach a condenser and heat the solution to reflux for 30 minutes.

5. Cool the solution to room temperature, then add 50 mL H_2O.

6. Pour the solution into a separatory funnel and drain the water layer.

7. Dry the organic layer with Na_2SO_4.

8. Pour (decant) the organic layer into a 50 mL round bottom flask.

9. Set up a simple distillation and collect the ethyl acetate in a conical vial.

A sample notebook page for a synthetic experiment is shown above. This is how the directions should appear **before** you begin the experiment. As you complete a step place a check mark next to it to keep track of your progress. Be sure to record any important observations or changes in the experiment. At the beginning of a synthetic experiment you should always write out the full chemical reaction, record the molecular weights of the important compounds, as well as their densities if they are liquids. These make calculating the theoretical yield a relatively trivial task. For a technique experiment show all of the important details of the experimental setup. Note that, for example, although the first time you do an extraction experiment, you will write out the experiment in detail, in later experiments the line—the solution was extracted 3 times with 25 mL of ether—is quite sufficient.

Many products are to be turned in to your laboratory instructor; when this is to be done you will be provided with small ziplock bags. The bag's label is to have clearly written on it: notebook reference number, the compound's name, m.p. or b.p., the student's name and laboratory section. Do **NOT** include your student number, unless you are specifically asked for it. The notebook reference number consists of your last name or initials, the notebook number, the page number, and a letter identifying the particular compound on that page. For instance, JB-34-D is compound D described on page 34 of JB's notebook.

To summarize, make sure to thoroughly record everything pertaining to your experiment in your notebook while **you are in lab**.

Apparatus

The beakers, test tubes, and Erlenmeyer flasks in your drawer should be familiar to you from previous labs. There is, however, considerable equipment that will be new to you. In this section a brief description of the more important items will be given.

First, ground joint glassware. Your drawer contains several pieces of 14/10 glassware. The 14/10 means the joint is 14 mm wide at the top and 10 mm long. These are standard taper joints, which means that they have a taper of 1 in 10. In other words, the glass is angled such that for every 10 mm of length the width contracts by 1 mm. They are interchangeable, so any male 14/10 joint will fit into any female 14/10 joint, regardless of manufacturer or age. They are also quite expensive. The joint alone costs about $5.00 new.

Ground glass joints, if they are kept clean, do not normally need grease in order to be used properly. However, they should not be left assembled as the joints may freeze together. It is very important to use grease if the reaction involves a strong base; otherwise the joints **will** freeze together. Two small dabs 180° apart **on the male joint** is sufficient amount of grease. Make the desired connection and rotate the male joint in the female joint to spread the grease over the entire mating surface. Excess grease should be wiped off with a tissue. When the apparatus is taken apart the grease may be removed with acetone.

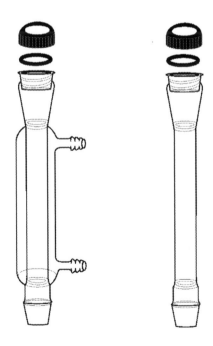

Figure I. West condenser on left, air condenser on right.

Now that you know what a 14/10 joint is, we will discuss what the various pieces of equipment are actually for.

1. *Round bottom flasks.* In this course, we use 10 and 25 mL round bottom flasks. These are what most of your reactions will be done in. Check carefully to make sure they have no cracks of any sort, they could cause a disaster if they break upon heating or under vacuum. Also remember they are *round bottom* flasks, and will roll around on a bench top. Cork rings are used for supporting round bottom flasks on the bench top.

2. Two different, but very similar pieces of equipment, are the *West condenser* and the *air condenser* (Figure I). They can be told apart by observing that the air condenser has no jacket while the West condenser has a water jacket. An air condenser can be used for small amounts of lower boiling point solvents or larger amounts of higher boiling solvents, while a West condenser can be used to condense even low boiling solvents in reasonable quantities. The downside is that the water jacket is much more expensive to make.

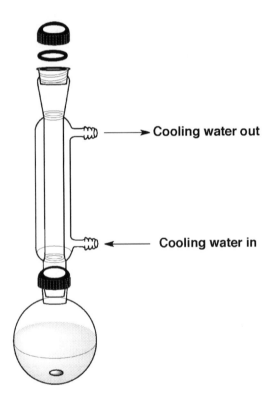

Cooling water out

Cooling water in

Figure II. How to connect the water hose.

When using the West condenser always make sure cooling water goes in the bottom tubulation and out to the drain through the top tubulation (Figure **III**. Always make sure your cooling hoses are clipped on to the tubulations, this is what the hose clips are provided for).

3. The *Claisen adapter* allows two pieces of apparatus to be connected simultaneously to a single necked round bottom flask (Figure IIIA). The distillation adapter allows the use of a thermoprobe to monitor a distillation while angling the vapors toward the condenser (Figure IIIB). The vacuum adapter allows a vacuum distillation to be performed as well as transferring distillate smoothly from the condenser to a receiver flask (Figure IIIC).

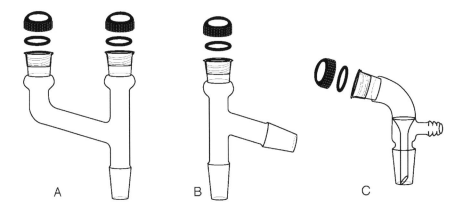

Figure III. "**A**" Claisen adapter, "**B**" stillhead adapter, and "**C**" vacuum adapter.

4. The *thermometer adapter* has many uses (Figure IV). Aside from its principal use in supporting the thermoprobe in ground joint setups, it can be used to hold any tube or rod of approximately the same 7–8 mm diameter. This includes, but is not limited to, inlet tubes, drying tubes, and outlet tubes. There is a small (but expensive) O-ring that goes over the item to be held (whether a thermoprobe or a tube) and just under the screw cap. Different manufacturers use different threads on their adapters, some are external, some internal. Be sure your screw cap fits and tightens on the glass piece you have. If it does not take your glass piece to the stockroom to obtain a screw cap that fits yours properly. The small O-ring is the same, regardless of manufacturer.

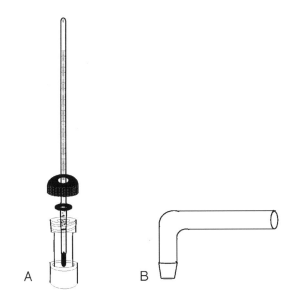

Figure IV. "**A**" Thermometer adapter, "**B**" Drying tube.

5. The *drying tube* is used to keep water vapor away from the surface of a reaction (Figure IVB). To use it, a small tuft of cotton is first inserted from the top and packed down to the indentation. Then one inch of $CaCl_2$ drying agent is added and an additional cotton plug is placed to hold the $CaCl_2$ in. Make sure the $CaCl_2$ and cotton are emptied out at the end of the period; otherwise, it tends to harden in place, which makes its removal very difficult.

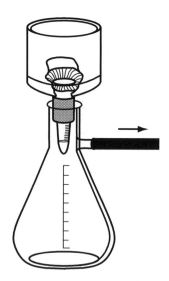

Figure V.   Filter flask with Büchner funnel.

6. The filter flask (Figure V) is used to perform vacuum filtrations. The trap adapter set vacuum hose is hooked up to the tubulation on the filter flask, and then the stopper end is inserted into an inverted filter flask, which serves as a trap. The aspirator vacuum tubing is then connected to the trap flask's tubulation. The appropriately sized rubber adapter is fitted to the neck of the filter flask, and finally the filter funnel assembly is firmly seated into the adapter. Always clamp the filter flask when you are using it, since the overall assembly is top heavy and unstable—and also they are expensive. Make sure there is an appropriate size piece of filter paper on the perforated plate. Büchner funnels have straight sides so it should be obvious that their filter paper lays flat on the plate. However, even though a Hirsch funnel has angled sides, this does not mean that the paper is folded to match the sides. A small disk of filter paper that just covers the perforations lays flat on the bottom of the funnel.

Always use a trap on your vacuum filtration setups if there is any possibility that you may need the filtrate later. If the aspirator backs up, there will be water in the filtrate that potentially could ruin it. Using your trap adapter set, hook up a trap between the aspirator and your filter flask. When using the aspirator always turn the water fully on. Using partial flow encourages backups.

The remaining apparatus in your drawer will be described as they are needed.

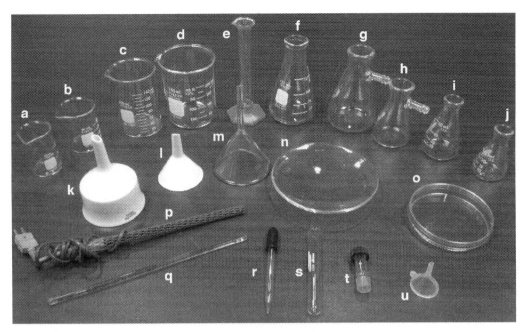

Figure VI Contents of Upper Drawers (Odd Number): a. 30 mL Beaker; b. 50 mL Beaker; c. 150 mL Beaker; d. 250 mL Beaker; e. 10 mL Graduated Cylinder; f. 125 mL Erlenmeyer Flask; g. 125 mL Filter Flask; h. 50 mL Filter Flask; i. 50 mL Erlenmeyer Flask; j. 25 mL Erlenmeyer Flask; k. Buchner Funnel; l. Hirsch Funnel; m. Glass Funnel; n. Watch Glass; o. Petri Dish; p. Thermoprobe; q. Stirring Rod; r. Medicine Dropper; s. Test Tube; t. Thermometer Adapter; u. Plastic Microfunnel.

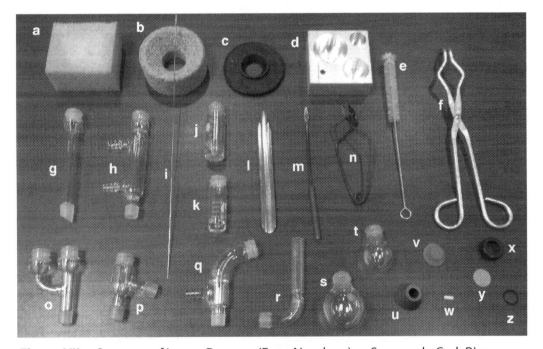

Figure VII Contents of Lower Drawers (Even Numbers): a. Sponge; b. Cork Ring; c. Filtervac; d. Aluminum Heating Block; e. Test Tube Brush; f. Tongs Crucible; g. Air Condenser; h. Jacketed Condenser; i. Stainless Steel Wire; j. 8 mL Conical Vial; k. 5 mL Conical Vial; l. Scoopula; m. Micro Spatula; n. Test Tube Holder; o. Claisen Adapter; p. Stillhead Adapter; q. Vacuum Adapter; r. Drying Tube; s. 25 mL Round Bottom Flask; t. 10 mL Round Bottom Flask; u. Filter Cone; v. Yellow Stopper; w. Micro Stir Bar; x. Round Hole Cap; y. Teflon Faced Septum; z. O-ring.

Laboratory Techniques— The 'How-To' Section

A. Melting Point Determination

Melting points are important physical characteristics of chemical substances. You will be using a melting point apparatus (Mel-Temp II, Figure 2.2), in order to collect melting point of your product. The Mel-Temp has a heating block with a slot for a thermoprobe, three grooves for capillary sample tubes containing your compound, a lighted observation window, and a knob that allows you to control the rate of heating. Follow the steps below to determine the melting point of your sample.

1. Carefully insert your stainless steel thermoprobe (not one coated with Teflon!) into the quarter inch sized opening.

2. Insert the yellow plug into the readout, making sure the polarized prongs are in the appropriate slot (one side is larger than the other).

3. Turn the readout on, be sure it is reading in °C not °F.

4. Insert the packed capillary sample tube (no more than 1 mm of material) in to one of the slots in front of the observation window.

5. Make sure the voltage control is set to zero and the on-off switch is in the "ON" position. You should be able to observe your sample through the observation window.

6. If you have no idea what your melting point is, set the voltage control at some intermediate level (maybe 5) and wait until you overshoot the melting point. This will be your rough-and-dirty determination. In order to determine the melting point, you need the temperature to be changing at only 2°C per minute—but if your melting point is 250°C, you do not want to take an hour to get close. Make sure your capillary gets into the glass waste when you are done.

7. Once you have some idea of what your melting point is, set the voltage so that you get within about 20°C at a reasonable pace and then turn the voltage down so that the rate of temperature change is the recommended 2°C per minute.

8. Record the temperatures at which the crystals first start to melt, and at which they finish melting. This is your melting point range. If the beginning and ending temperature appear to be the same, your temperature may have been changing too quickly; or else you have a very pure sample. Impure samples tend to have fairly wide temperature ranges that are usually lower than the melting point of the pure sample.

B. Recrystallization

Recrystallization is a technique used to purify compounds. Some type of purification is necessary just about any time you isolate or synthesize a chemical. Recrystallization involves dissolving your crude and impure chemical mixture in a hot solvent, and slowly crystallizing the desired compound while leaving the undesired impurities in solution. Impurities are removed either by filtration (if they did not dissolve during a hot filtration) or remain in solution after the product has crystallized if they are more soluble even in the cold solvent. The following guidelines should help you perform a recrystallization.

1. Heat some of the solvent you plan to use either on a hot plate or steam bath. How much is "some"? That depends on how much solid you have, so check with your instructor if you are not sure. Feel free to pass any leftover solid on to another student to cut down on waste.

2. Put the solid to be purified in an appropriately sized round bottom flask and place the air condenser on top (be sure to leave the top open, never heat a closed system!).

3. Add the hot solvent until the solid just dissolved and then add small excess so that it will stay dissolved. If your solvent cools down too quickly you might heat the flask slightly while you are adding the solvent. Make sure not to keep adding solvent if there is only some gunk left. You do not have to dissolve all of the impurities; you can filter them out.

4. If you have any solid remaining that does not appear to be your product, you will want to filter them out before you allow your solid to recrystallize. For most of your work, you will need to use the small Hirsch funnel.

5. Allow the solution to cool slowly. The more slowly the crystals form, the less likely you are to end up with impurities trapped in them. Once the flask has cooled to room temperature, you may place it in an ice bath to cool further. Be careful that your flask does not tip over in the ice bath.

6. If your flask is cool, and no crystals have formed, it may help to scratch the inside bottom part of your flask with a glass-stirring rod. Often crystals grow outward from your scratch line.

7. Filter your crystals using the Hirsch funnel. Spread them out on the filter paper and allow air to be pulled through them to help them dry more quickly.

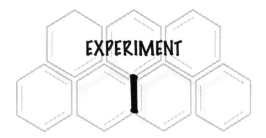

EXPERIMENT 1

Paper Chromatography

Introduction

Chromatography is a method of separation of compounds based on the principle of phase distribution. There are several different kinds of chromatography: thin layer, column, gas, paper, and liquid chromatography are among the most common types used in organic chemistry. All methods of chromatography involve the partitioning of a substance between a stationary phase and a mobile phase. In paper chromatography the paper itself serves as the stationary phase and the eluting solvent as the mobile phase. You will perform a paper chromatography experiment, which uses paper (coated with water from the air) as the stationary phase, and a 2-propanol/water mixture as the mobile phase. The liquid will run along the paper and will carry compatible compounds very quickly while incompatible compounds will move at a slower pace. You will be looking at colored compounds (dyes, inks) so that you may easily tell how far each component of the mixture has traveled.

Experimental Procedure

1. Determine the center of a piece of 12.5-cm filter paper by folding it in quarters.

2. Use this as a guide to poke holes in two other pieces of filter paper.

3. Use a capillary tube to drop a dime-sized spot of your assigned dye in the center of 1 filter paper. Allow this to dry.

4. On your other filter paper, use an ink pen to draw a similarly sized spot.

5. Tear off two 1 × 2 cm pieces from your folded filter paper and roll them into wicks 1 cm long. These should fit snugly in your two remaining pieces of filter paper, with about 8 mm extending on the bottom side.

6. Put approximately 30 mL of the eluent solvent (70:30 isopropyl alcohol:water) into the bottom halves of two Petri dishes, set a filter paper on each one (make sure the end of the wick dips into the solvent), and cover with the tops of the Petri dishes.

7. Let the chromatograms develop until the solvent is a few centimeters from the edge of the filter paper (at least 10 minutes).

8. Remove the chromatograms from the Petri dishes and allow them to dry.

9. Dispose of your solvent as directed by your instructor and compare your results with other students.

Laboratory Safety

The chemicals you will work with in this laboratory are less toxic than many you will encounter later this semester. However, 2-propanol is flammable and absorbs relatively easily through the skin. Make sure to wear goggles to protect your eyes from splashes.

Questions: _____ Name: _____ PID: _____

1. How many components are there in the green dye?

2. What are the colors of the components of the green dye?

  a. Which of the components of the green dye is the most polar?

  b. What color are the components of the black ink?

  c. What is the stationary phase in this experiment?

  d. What is the mobile phase in this experiment?

3. Draw the components of the black ink as they appear on the paper. Which of the component of the black ink is the least polar?

EXPERIMENT 2

Melting Point

PART 1

Thermometer Calibration TC "Melting Points"

Everyone should calibrate a digital thermometer. Make sure to write down the digital thermometer number in your lab notebook. Use the same digital thermometer every time you need to use the thermoprobe (also called thermocouple). This is not a group project. Accurate determination of melting points helps your grade significantly since a portion of your score in the practical experiments depends on accurate reports of melting and boiling points. If you break or lose your thermoprobe, get a new one from the stockroom and repeat the calibration procedure.

Pure crystalline solids usually melt over a narrow temperature range. The intermolecular forces in organic crystals are relatively weak Van der Waals or hydrogen bonds, and the melting points for organic compounds are much lower than most common inorganic compounds, which tend to have ionic bonds. The easy accessibility of the melting range of organic substances makes their melting point a useful physical property for their identification.

A mixture of two pure compounds has a lower melting point than either single component but a larger melting point range. Thus, mixing two components with identical melting points gives a product with a lower melting point. This property can be used to estimate the purity of a compound (impure organic compounds melt over broad ranges) or confirm the identity of a material. Identification can be confirmed by mixing the unknown compound with a pure sample of the suspected compound and taking the melting range.

While taking a melting point range, the first report is made when the crystals begin to melt and the final when all of the crystals have melted. Although some handbooks and manuals list a single melting point for particular compounds, this is a misnomer: they are listing only the beginning of the melting range. Make sure when you record a melting range you list both numbers.

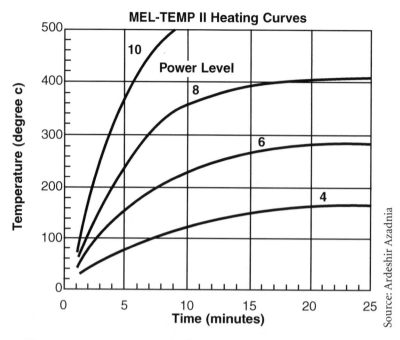

Figure 2.1 The power-setting guide for Mel-Temp II melting point apparatus.

We have provided purified samples of naphthalene (80.2°C), benzoic acid (122.24°C), and salicylic acid (158.3°C) for the calibration of the digital thermometers. Remove the stainless steel thermoprobe from its slot in a melting point apparatus and immerse it into an ice/water bath in order to check the 0° reading and then determine the observed melting points of the above three compounds using the same digital thermometer and thermoprobe. Write down the number on the digital thermometer and the Mel-Temp and always use the same ones. Plot the true (vertical) vs. observed (horizontal) data in your notebook. Use the plot to determine any corrections that need to be made for reporting melting points or boiling points of products. Determine the optimum sample size by melting 1 crystal of benzoic acid, a 0.5-mm stack of crystals, and a 5–mm sample simultaneously.

Although there are many ways to obtain melting points we will only describe the one, which you will be using in CEM 143. By following these steps a good melting range may be obtained.

1. Obtain a melting point capillary tube. These have one end sealed.

2. Tamp the open end of the tube into a small pile of the compound whose melting point you plan to measure.

3. By tapping the closed end gently on a book, the compound will be forced to the bottom of the tube. Only 0.5 mm of compound is needed.

4. Find an unused melting point apparatus equipped with a digital thermometer (Figure 2.2) and write down the number of the apparatus. Each of the melting point apparatuses are numbered and have a digital thermometer and thermoprobe assigned to them. You should remember to always use the same melting point apparatus in the future. Do not remove the thermoprobe from the melting point apparatus. You have one in your own drawer. Make sure the heater is turned off. Ensure that the current apparatus temperature will not immediately

melt your compound, if it is too hot either use another apparatus or wait until the heater block is cool enough to use.

Do NOT cool the melting point apparatus with water.

5. Place your three samples in the appropriate slots and adjust the power control to a setting appropriate for the expected melting point of your compound (see the Mel-Temp II setting guide), (Figure 2.1). Start with power setting of about "3" (for naphthalene), increase it to "3.5" (for benzoic acid), and finally after both benzoic acid and naphthalene have melted, move up the power setting to "4." Do not ever turn the power setting above "6." Heating too rapidly will result in bad data while going too slow results in wasted time.

6. Record (in your laboratory notebook not on a scrap of paper!) the temperatures at which the compound first begins to melt and when the entire sample has melted. This is called the melting point range of a compound.

7. Turn the apparatus off, place the melting point apparatus back on the shelf, and dispose of the melting point capillary in the Broken Glassware Glass Container.

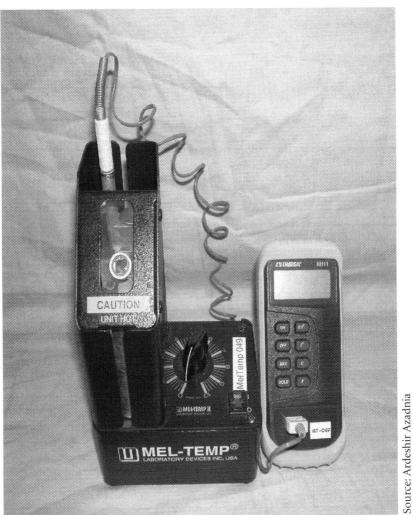

Source: Ardeshir Azadnia

Figure 2.2   Mel-Temp II° Melting Point Apparatus.

This is also the best time to wash all your glassware. Warm water, soap, a brush, and lots of elbow grease will remove most dirt. Resistant organic "crud" may be removed by rinsing with a bit of technical grade acetone—which is available at the stockroom window. Follow it with soap and water. Stubborn inorganic deposits may be dislodged with acid. Check with your instructor or the stockroom. Rinsing thoroughly with water and finally distilled water should allow the glassware to dry during the week without leaving any deposits or watermarks. Washing any dirty glassware before you leave saves loads of time in the following week. No need to mess around trying to dry those nooks and crannies. Do it every week before you leave. If there is a pause in the experiment, do some dishes. Clean up and lock up before you leave each week.

PART II

You will be expected to discuss the physical properties of the organic compound(s) found on the product label during the consumer product poster session. You should always report the melting point range in your lab report. For graphing of the calibration curve, take the average of the melting points by adding the two numbers and dividing by two.

Experimental Procedure

1. Your instructor will assign you and your partner an unknown compound. You are to determine an accurate melting point for your unknown and report it. Frequently, students lose many points due to poorly measured melting points.

2. Use your own thermoprobe to measure the melting point of ice in distilled water first. Then dry the thermoprobe and measure the melting points of naphthalene (80.2°C), benzoic acid (122.24°C), and salicylic acid (158.3°C).

3. Prepare a calibration curve by plotting your observed melting point values (X-axis) against the literature values (Y-axis).

4. Record the number on the melting point apparatus in your lab report. For better accuracy, you should always try to use the same melting point apparatus (identified by their unique numbers) when taking melting points.

5. Both you and your laboratory partner should perform a melting point determination of your unknown sample—the first to go will do the "rough-and-dirty" in order to find the approximate melting point range, and the second partner will do a more careful one.

Note: You should always remember to use your calibration curve to correct the observed melting point during the later experiments and report them as "corrected melting point."

Laboratory Safety

Some of the compounds you will work with are flammable, toxic, and absorb through skin. If you spill any chemicals on yourself rinse thoroughly (for 20 minutes) with cold water, then with soap and water, and finally with water again while another student (your lab partner) should notify the lab instructor immediately.

Questions: _____ Name: _____ PID: _____

1. What is the melting point range of your unknown compound?

   First try:

   Second try:

2. Suppose, due to an error, a student heats the melting point apparatus at high power settings such as 7 or 8!

   Explain how this may affect the observed melting point of his unknown. Would the melting point of his unknown be higher than the true value or lower?

   Explain your reasoning.

3. Graph a calibration curve for the thermoprobe of the melting point apparatus and turn it in with your lab report.

EXPERIMENT 3

Distillation of Ethanol/Acetone

Background and Theory

Distillation is the simplest and one of the most common ways to purify a liquid substance. It involves heating the liquid until it boils and then condensing the vapor to liquid away from the original solution. One cycle of vaporization/condensation constitutes a simple distillation. This works very well as a purification method if the impurities in the substance have no significant vapor pressure at the boiling point of the liquid. If the impurities do have a significant vapor pressure at the liquid boiling point then a simple distillation will not succeed in giving the pure liquid. The distillate will, however, be enriched in the lower boiling compound. Further simple distillations will progressively improve the purity of the distillate, but at the expense of much time and effort. Thus, the greater the boiling point difference, the easier the separation of two components by distillation.

To solve this problem, we take advantage of the fact that multiple distillations done in the same apparatus would be the equivalent of multiple sequential simple distillations. This process is called fractional distillation. Use of a column increases the distance that vapors from the distillation flask have to travel to reach the condenser. Hence, one can accomplish multiple simple distillations in a single reaction vessel because of the extra surface area of the column. All else being equal, a longer column gives a better separation as the condensation/evaporation process is repeated more times. Adding a packing material to the column, such as glass beads, glass helices, or metal sponge, also increases the efficiency of the distillation.

The relative efficiency of a distillation apparatus is measured in terms of theoretical plates. One theoretical plate is equivalent to one simple distillation; thus higher number of theoretical plates in an apparatus will result in a better separation. Column efficiency can also be measured in terms of theoretical plates per unit length (i.e., theoretical plates/cm). This is calculated by dividing the column length by the number of theoretical plates in that column.

The Fenske equation can be used to calculate the number of theoretical plates in a particular piece of apparatus. For a simple two compound system the Fenske equation can be written as:

$$n = \frac{\log\left(\frac{X_a}{X_b}\right) - \log\left(\frac{Y_a}{Y_b}\right)}{\log \alpha}$$

n = number of theoretical plates

X_a = the vapor pressure of the most volatile compound in the product (**acetone**).

X_b = the vapor pressure of the most volatile compound in the distilling flask (**acetone**).

Y_a = the vapor pressure of the least volatile compound in the product (**ethanol**).

Y_b = the vapor pressure of the least volatile compound in the distilling flask (**ethanol**).

α = vapor pressure ratio of the two components.

The percentages of each of the components in both the product (X_a & Y_a) and that remaining in the distilling flask can be best determined by using gas chromatography technique (GC). The vapor pressure ratio α, may either be looked up in tables for common compounds, or determined experimentally from the distillate of a distillation known to have only one theoretical plate.

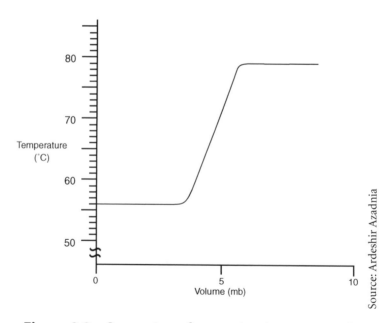

Source: Ardeshir Azadnia

Figure 3.2 Separation of an equimolar mixture of acetone/ethanol by distillation. The boiling points for the pure components are 56 & 78.5°C for acetone and ethanol, respectively.

Experimental Procedure

Simple Distillation

1. Assemble a simple distillation apparatus as shown in Figure 3.2. Have your lab instructor check your setup before you start. Use a 25–mL round bottom flask as the distilling flask and a 10–mL graduated cylinder to collect the distillate.

2. Carefully, remove the thermoprobe/thermoprobe adapter and introduce 14 mL of equimolar acetone/ethanol mixture using a clean short-stemmed funnel. Add a pair of boiling chips to the 25–mL round bottom flask before replacing the thermoprobe.

3. Turn on the water to the West condenser gently, and start heating. **Do not forget the boiling chips!**

4. Record the boiling point after every 0.4 mL increase in distillate volume. If you observe a rapid temperature change, collect more data points.

5. Be sure that you have enough data points to draw a good graph for your distillation experiment. Plot this data (volume = horizontal axis, and temperature = vertical axis) in your notebook while performing the experiment (Figure 3.1).

6. Stop when the residue in the distilling flask is reduced to about 3 mL.

The most commonly asked question at this point is "What is the correct setting for the hot plate power controller?" The answer is "It depends." The amount of liquid being distilled, heat capacity, boiling point, and heat of vaporization of the liquid, the size of the flask, the initial temperature, the electrical characteristics of the hot plates, and the rate of boiling all affect the "right" setting. The settings must be determined by experiment and adjusted as conditions warrant it. Think of the power controller as an accelerator pedal. Use a high setting (about 7 or 8) when first starting, then reduce the setting as bubbles begin to form. The best separation will occur when the collection of the liquid in the 10–mL-graduated cylinder is one drop per one second. If you "over-accelerate" and the rate of distillation exceeds this, turn off the controller and lower the aluminum-heating block for a few minutes to allow the excess heat in the block to dissipate.

Report your data as below:

Simple Distillation		**Fractional Distillation**	
Volume (mL)	Temp. (°C)	Volume (mL)	Temp. (°C)
————	————	————	————
————	————	————	————
————	————	————	————
————	————	————	————
————	————	————	————
————	————	————	————
————	————	————	————
————	————	————	————
————	————	————	————
————	————	————	————
————	————	————	————
————	————	————	————
————	————	————	————
————	————	————	————
————	————	————	————
————	————	————	————
————	————	————	————
————	————	————	————
————	————	————	————
————	————	————	————
————	————	————	————
————	————	————	————
————	————	————	————
————	————	————	————
————	————	————	————
————	————	————	————
————	————	————	————
————	————	————	————
————	————	————	————
————	————	————	————
————	————	————	————
————	————	————	————

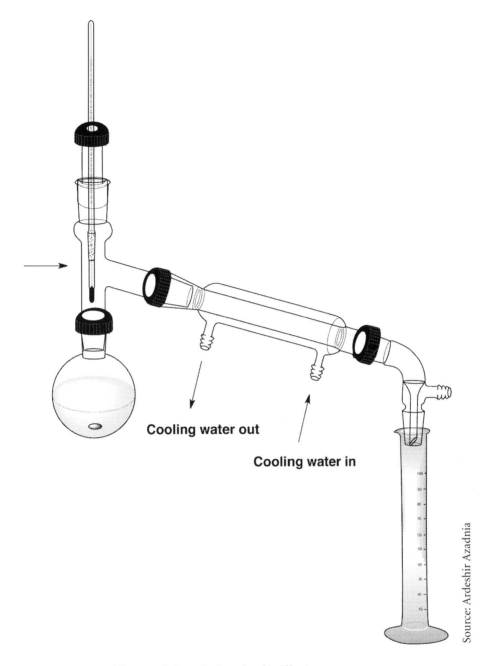

Cooling water out

Cooling water in

Source: Ardeshir Azadnia

Figure 3.2   A simple distillation setup.

Fractional Distillation:

7. Cool the apparatus, place 14 mL of the equimolar acetone/ethanol mixture to a clean 25–mL round bottom flask.

8. Attach a Claisen adapter and an air condenser to the flask, as shown in Figure 3.3. Reassemble the rest of the distillation apparatus, and repeat the distillation process.

9. Heat at a rate of one drop per second. The heating rate will have to be adjusted as the volume changes.

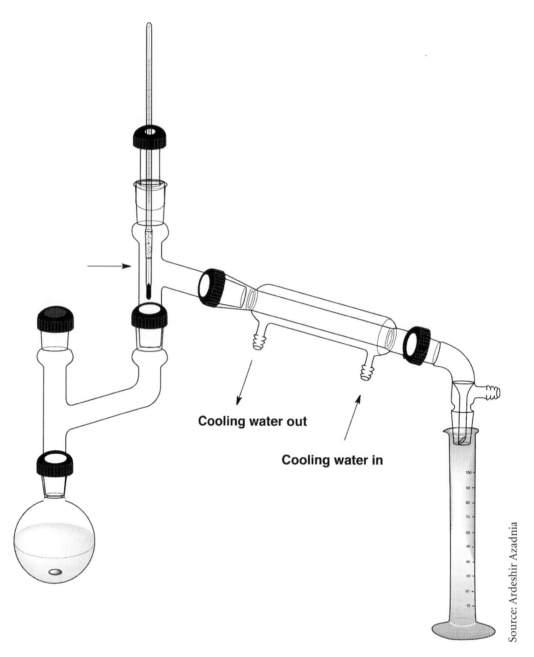

Figure 3.3 A fractional distillation setup.

Cooling water out

Cooling water in

Source: Ardeshir Azadnia

11. Plot the data on the same graph of the simple distillation.

12. The Claisen adapter functions as a fractionating column. Try to regulate the rate of heating so that a distinct separation of acetone and ethanol is observed.

13. Use this graph to estimate the volumes of acetone and ethanol in the mixture. Which apparatus gave a better separation? Why?

At the end of the period, the distillate and any residue should be poured into the Hazardous Waste Container; it should not be poured down the drain. You need not wash the apparatus since traces of both acetone and ethanol will evaporate without a trace.

Questions: _____ Name: _____ PID: _____

1. What is the definition of simple distillation and theoretical plates?

2. Think of two ways to increase the number of theoretical plates in a distillation setup.

3. A student in CEM143 was doing a fractional distillation when she noticed that the temperature suddenly started to drop. What could have caused the decrease in the temperature? Provide two possible reasons for the above.

4. A student in an organic chemistry lab was purifying xylenes (boiling point = 130°C). However, by mistake, he had placed his thermoprobe above the arm of the still head adapter. Explain how this would affect the temperature readings for the boiling point of xylenes. Would the boiling point reading be higher or lower than the actual one? Explain your answer.

EXPERIMENT 4

Acid-Base Extraction

EXTRACTION

Background and Theory

Liquid-liquid extraction is the most efficient and economical technique for separating carboxylic acids and amines. For a successful acid-base extraction, two immiscible solvents must be used. For instance, oil and water are two immiscible liquids. Also, dichloromethane (methylene chloride) and water are two immiscible solvents. If they are mixed, they will quickly separate into two layers. However, if a compound, which is at least partially soluble in both solvents, is added before shaking, then this compound will be present in both layers after they settle. The concentration in each layer depends upon the attractive and repulsive forces between each solvent and the solute. For a given compound and a pair of solvents, the ratio of concentrations is a constant, K, known as the "distribution coefficient" or "partition coefficient." Note that this is in terms of concentration (e.g., g/mL) and not total amount.

$$K = \frac{g[A]/mL[\text{org}]}{g[A]/mL[H_2O]} \quad \text{or} \quad \frac{\text{molarity}[\text{org}]}{\text{molarity}[H_2O]}$$

The preceding discussion is based on the assumption that the compound does not react with the solvent system. For systems which do react, the theory still works, provided the distribution coefficient for the proper compound is considered. If a mixture of 3-nitrobenzoic acid (the acid) and methyl 3-nitrobenzoate (the ester) is extracted by ether/water, most of the acid and the neutral compound will be found in the organic layer as most organic compounds are insoluble in water. However, if the same mixture above is extracted with ether and 0.25 M aqueous sodium carbonate, the acid will react with the hydroxide ion to form the water–soluble sodium 3-nitrobenzoate and end up in the water layer while the ester will end up in the ether layer as shown in Scheme 5.1. As we know, all sodium, potassium, and ammonium salts are water-soluble while they are virtually insoluble in organic solvents. On the other hand most carboxylic acids are insoluble in water. This principle will be used in this experiment to separate an organic acid from an ester (neutral compound).

Reaction of the sodium 3-nitrobenzoate with dilute H_2SO_4 will result in the water insoluble 3-nitrobenzoic acid. In this experiment, we will take advantage of this acid–base reaction to separate 3-nitrobenzoic acid from methyl 3-nitrobenzoate.

Scheme 4.1   Separation of a carboxylic acid from an ester.

The ester is insoluble in aqueous solution regardless of pH, and will remain in the organic solvent (ether layer) while the acid is dissolved in a basic aqueous solution. Separation of the layers in a separatory funnel or a conical vial allows one to isolate the compounds.

The final products isolated in this experiment are solids. Therefore, you will need to learn how to purify them. Recrystallization is the simplest and most economical method for purification of organic solids.

Recrystallization

In recrystallization, the impure solids are dissolved in minimum amount of a suitable hot (boiling) solvent so that when the solution is cooled to room temperature, the impurities would stay dissolved and the solid of interest crystallizes out of solution. This technique takes advantage of the fact that as a crystal grows in a solvent system it tends to add only molecules of its own kind to the crystal, resulting in a solid of high purity. The first step in recrystallization is to find a solvent in which the compound is highly soluble in when hot, but is nearly insoluble when cold. In addition, any impurities should remain soluble in the cold solvent. Once a suitable solvent is found the compound is dissolved in a minimum amount of hot solvent. The flask is then left undisturbed until crystals have formed. It is a good idea to cool the flask containing the recrystallized product in order to maximize the yield. The solids are collected by vacuum filtration (Figure 4.1). One must not forget that most compounds have a small but finite solubility in cold solvents. Therefore, we must use the absolute minimum amount of solvent required to dissolve the solid initially. There are two principle complications in recrystallizations; lack of a single suitable solvent and insoluble impurities.

Using solvent pairs often solves the first difficulty. Such solvents should have the following characteristics:

1. They should be completely miscible.

2. One solvent should be a good solvent for the compound to be recrystallized, and the other has to be a poor solvent for it. For example: ethanol/water, methanol/water, ethyl acetate/hexanes, and dichloromethane/hexanes are commonly used solvent pairs used for recrystallization of organic solids. Many organic compounds are too soluble in ethanol for it to be used alone for recrystallization. Since most of these compounds are insoluble in water, and water and ethanol are completely miscible, they form a suitable solvent pair. For instance, the compound is dissolved in the minimum amount of a hot ethanol. Then water is added dropwise, while the mixture is still hot, until the solution turns cloudy. Then, one drop of the ethanol is added and the solution is then set aside to cool, slowly. The key word in the recrystallization technique is "slowly" getting the crystals to come out of solution. The slower the recrystallization process, the more pure the compound would be.

The second difficulty in recrystallizations is insoluble impurities. These are removed by hot filtration. Once the compound is completely dissolved in a hot solvent it is filtered through a fluted filter paper in a stemless funnel. Both the stemless funnel and the fluting of the filter paper serve to prevent crystals from forming during the filtration. After filtration, the solution is allowed to cool and the product (hopefully) crystallizes. Decolorizing carbon, added to adsorb trace quantities of highly colored impurities, is also removed by a hot filtration.

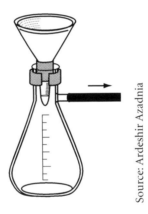

Source: Ardeshir Azadnia

Figure 4.1 Vacuum filtration setup.

Both the nature of the compound and the rate of cooling influence crystal size. The slower the rate of cooling, the larger the crystals. Up to a point (rarely seen in introductory organic lab) the larger crystals will tend to be more pure. Therefore, cool your crystals slowly. Most of your recrystallizations will be done on a steam bath. Once the compound is completely dissolved, cool the solution on the bench top first, and then, once it reaches room temperature, put it in an ice bath.

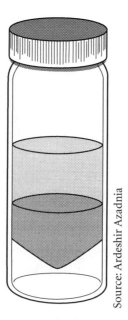

Source: Ardeshir Azadnia

Figure 4.2   8–mL conical vial.

CAUTION: You will be working with many HIGHLY FLAMMABLE liquids. Ether is extremely flammable. Ether vapors will ignite if they come in contact with hot surfaces of 200°C or higher.

Introduction. Both components of the extraction mixture are solids. **Save and label all solutions** until you have isolated both solids. Extraction techniques tend to be messy and smelly. Do not oven dry the solids.

Experimental Procedure

I. Separating the Acid from the Neutral Compound

a) Weigh out 400 mg (0.4 g) of the solid extraction mixture and thoroughly pulverize the solid on a piece of smooth weighing paper.

b) Combine the powdered solid, 4 mL of diethyl ether (a.k.a. ether—HIGHLY FLAMMABLE), and 2 mL of 0.25 M sodium carbonate in an 8 mL conical vial (Figure 4.2).

c) Vigorous swirling should dissolve the solid (use the automatic mixer for efficient mixing).

d) Remove the aqueous layer as instructed below:

e) Place a small piece of cotton into a 5-1/2-inch Pasteur pipette and insert it all the way inside the tip using the special stiff wiring provided in your drawer (your TA will demonstrate this for you).

f) Press the bulb of the Pasteur pipette and insert it into the bottom of the conical vial and draw the aqueous layer slowly into the pipette.

g) Displace the aqueous layer into a 50 mL beaker. Repeat if needed until all aqueous layer is removed.

h) Extract the ether layer with another 2 mL of 0.25 $\underline{M}$ sodium carbonate solution. Shake the mixture thoroughly.

i) Allow the layers to separate and remove the aqueous layer and pour it into the 50 mL beaker. Keep the emulsion layer with the ether.

j) The beaker now holds all of the organic acid, in solution as its conjugate salt, RCO_2^-, Na^+. Save the ether layer containing the neutral compound for Part III.

II. Isolation of the Neutral Compound

1. No matter what organic solvent is used in an extraction, one needs to use a **drying agent** to get rid of the dissolved water in the organic solvent. (Ether can dissolve 1.2 percent water at 20°C). This is known as "drying" the organic layer. Anhydrous magnesium sulfate and sodium sulfate are two of the most common drying agents used in organic laboratories.

a) Prepare a micro-drying column as follows (Figure 4.3).

b) Place a small piece of cotton into a 5-1/2 inch Pasteur pipette loosely.

c) Using your micro funnel, add 4 mm sand and 2 cm of anhydrous sodium sulfate.

d) Transfer the ether layer containing the neutral compound through the above sodium sulfate column into a 10 mL round bottom flask.

e) Rinse the conical vial with two 1 mL ether and pass through the sodium sulfate column.

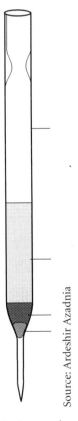

Source: Ardeshir Azadnia

Figure 4.3   Drying column.

2. Assemble a simple distillation apparatus and remove the ether. Make sure to use a boiling chip or your stirring bar to avoid violent bumping of the solution. The used sodium sulfate may be discarded.

3. Return the distilled ether to the hazardous waste container. The neutral compound will be a yellow oil at this point.

4. Dismantle the distillation setup and heat briefly with the Stillhead removed to eliminate the last few drops of ether. Failure to remove all of the ether interferes with the crystallization steps.

5. Add 1.5 mL of methanol to the oily distillation residue, mount your air condenser, and bring the mixture to a boil on a steam bath (Figure 4.4). The air condenser will prevent methanol from evaporation.

6. Continue heating and add water dropwise until a trace of permanent cloudiness is present. It may require up to 15 drops.

7. Add 1 or 2 drops of methanol to remove the cloudiness. Cloudiness here may also be caused by the residual ether.

8. Transfer the hot solution into a clean, dry, and warm 10 mL Erlenmeyer flask via a Pasteur pipette and set the solution aside to crystallize.

9. After 10 minutes of slow cooling, solid crystals should appear. The flask may be clamped in an ice water bath to complete the crystallization and maximize the yield.

10. If no solids are formed after 10 minutes, scratch the inside of the flask with a glass-stirring rod to induce crystallization.

11. Consult with your instructor if you are unsuccessful.

Source: Ardeshir Azadnia

Figure 4.4   Steam bath.

12. Set up a vacuum filtration (Figure 4.1) using a 50 mL filter flask and a Hirsch funnel and collect the crystalline neutral compound by suction filtration.

13. Take and report the melting point.

   Note:

   a) Secure your filter flask by clamping.

   b) Make sure to use a thick–walled tubing to connect the filter flask to an aspirator.

III. Isolation of the Organic Acid

1. If the carbonate layer from Part II is cloudy, filter it into a 50-mL Erlenmeyer flask through a pea-sized ball of loose cotton in the micro-funnel to remove the impurities. The impurities are coarse enough to be trapped by the cotton.

2. Place the 50-mL Erlenmeyer flask in an ice bath, add a magnetic stir bar and acidify the clarified sodium carbonate solution of the organic acid with 1.5 mL of $2\underline{M}$ H_2SO_4.

3. Add the acid slowly and with stirring, as the mixture will evolve CO_2 gas. Before you come to lab, calculate the theoretical mass of CO_2, which will be generated.

4. The well-mixed suspension should now turn blue litmus pink. If the solution is not yet acidic, add a little more acid, mix well, and test with litmus paper.

5. Organic acids precipitated this way form very fine particles and filter only very slowly. Do not bother trying to collect the solid at this point.

6. Instead it is easier to extract the organic acid back into ether. Extract the acidified water layer with two 3-mL portions of ether as before in a clean 8-mL conical vial.

7. Discard the water layer and combine the two 3-mL aliquots of ether extract. This combined ether layer now has all of the organic acid.

8. Dry the ether layer by passing it through a freshly packed sodium sulfate column into a 10-mL pear shaped flask.

9. Assemble a simple distillation apparatus, add a boiling chip, and remove the ether by heating over a steam bath. Do not use a hot plate.

10. Collect the used ether in a flask and dispose of it promptly in the hazardous waste container. Heat with only the minimum amount of steam or it will boil over.

11. When no more ether distills, remove the distillation apparatus and carefully break up the lump of solid with a stirring rod.

12. Carefully crush the residue in the flask, add a small amount of ethanol, and place an air condenser and heat via a steam bath until all solids have dissolved.

13. Slowly add water until the solution is cloudy. Then add one or two more drops of ethanol to turn the mixture clear.

14. Leave the flask undisturbed for about 10 minutes to crystallize.

15. Scratching may be needed to start crystal growth. Check with your instructor if you have a problem getting crystals.

16. Once crystallization has started, clamp the flask in an ice bath to complete the process. This crystallization is slow! Plan on icing for at least 5 minutes.

17. Collect the "free acid" by suction filtration (Hirsch). Use the clear filtrate (mother liquor) to rinse any remaining solids into the funnel.

18. Dry and package the product in a vial. The filtrate can be poured down the drain with running water.

IV. Determine the Melting Points

1. Take the melting point of the recrystallized acid, the recrystallized neutral compound, and the starting mixture. All three can be done at the same time—there are three holes for samples in each Mel-Temp®.

2. Turn in your labeled products and your report before leaving the lab.

3. Your report should include the percentages of each component recovered based on the original sample taken as well as the total percentage recovered.

Questions: _____ Name: _____ PID: _____

1. **(1 Point)** Caffeine has a distribution coefficient of 8.4 for methylene chloride/water. How much caffeine will be extracted from a 60 mL cup of Turkish coffee (100 mg caffeine) by 40 mL of methylene chloride? Hint: let $\times$ be the mg of caffeine in methylene chloride at equilibrium.

2. **(1 Point)** Is it absolutely necessary to have a pair of solvents that are immiscible in order to have a successful extraction? Why or why not?

3. (**1 Point**) Classify the following as miscible or immiscible mixtures.

   a. Cooking oil and water:

   b. Ethanol and water:

   c. Milk and water:

   d. Diethyl ether and water:

Preparation of Methyl 3-Nitrobenzoate

BACKGROUND AND THEORY

Preparation of Methyl 3-Nitrobenzoate

Carboxylic acids are common organic intermediates and products. They may be prepared in a number of different ways. Being almost fully oxidized, they are a rather robust functional group, surviving and being formed under conditions that would destroy many other functional groups. The mechanistic details of each reaction will be discussed just before the experimental details of a particular reaction are discussed.

The nitration of methyl benzoate with a mixture of concentrated nitric and sulfuric acids is an example of an electrophilic aromatic substitution. As the carbonyl of an ester is an electron-withdrawing group (a meta-director), the nitro group is added to the 3 position of methyl benzoate (Figure 5.1). It is of note in this reaction, that nitric acid functions as a base, not an acid. The sulfuric acid also removes water from the reaction mixture, providing an additional driving force for the reaction.

Figure 5.1 The synthesis of methyl 3–nitrobenzoate.

From a mechanistic viewpoint, the formation of the actual electrophile is quite interesting. **As noted before, nitric acid acts as a base, accepting a proton from the sulfuric acid (acting as an acid, Figure 5.2).** This intermediate then loses water to produce the nitronium cation, the actual electrophile for this reaction. The ester group, acting as a *meta*-director, stabilizes the addition of the nitronium cation at the 3-position, giving intermediate **II**. Removal of the proton by water, gives the final product, **III**.

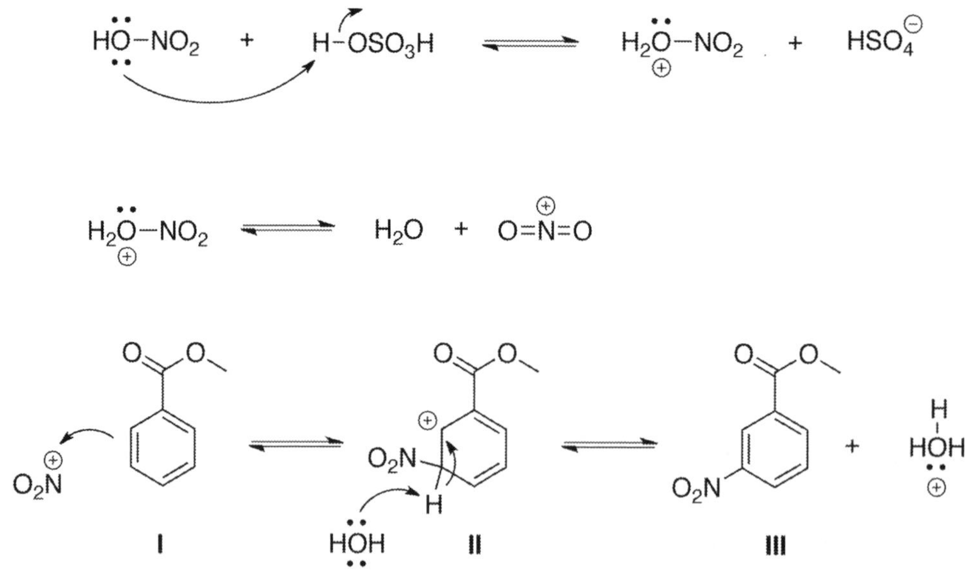

Figure 5.2 Mechanism of the nitration of methyl benzoate.

CAUTION

This experiment involves the use of concentrated nitric and sulfuric acids. Work carefully, rinse all apparatus immediately, and wipe up spills promptly. **First aid treatment** for acid spills on flesh is immediate and continued rinsing with water for 20 minutes. First aid for acids on clothing consists of immediate removal followed by generous rinsing. Let your lab instructor know promptly in either case.

Experimental Procedure

I. Nitration

1. All glassware must be **dry**. Excess water will stop the reaction. Pump 7.5 mL (three strokes from an automatic dispenser) of concentrated H_2SO_4 ($18\underline{M}$) into a 50 mL Erlenmeyer flask. Hold the flask so that the delivery tube is inside the flask while you operate the pump. The white powder on the tray is sodium carbonate; it will neutralize any acid which lands on it. Clamp the flask in an ice bath to cool the acid.

2. Pump 2.5 mL of conc. H_2SO_4 and 1.75 mL of concentrated HNO_3 ($16\underline{M}$) into a 25 mL Erlenmeyer flask and clamp it in the ice bath as well.

3. Pour the premeasured 2.1 mL sample of methyl benzoate (d = 1.09 g/mL) into the **chilled** 50-mL Erlenmeyer flask containing H_2SO_4. Swirl to mix and cool for 5 minutes.

4. Place your blue magnetic stir bar into the 50-mL Erlenmeyer flask and clamp the flask in an ice bath and start stirring (via a magnetic stirrer).

5. Transfer the nitric acid solution from the small flask to the stirred methyl benzoate solution (larger flask) dropwise with an eyedropper. Keep the eyedropper upright; if the

acid mixture drains into the rubber bulb, the product will be contaminated. When not in use, the eyedropper may be left standing in the nitric acid flask.

6. Transfer of the nitric acid solution should take about 2–5 minutes. Once the addition is complete, remove the flask from the ice bath and allow the solution to warm to room temperature for 10 minutes while stirring.

II. Product Isolation

1. Put about 25 g (50 mL, loosely packed) of ice in a 250 mL beaker.

2. Pour the acid mixture slowly over the ice while stirring with a glass rod.

3. Rinse any residues from the flask into the mixture with a small amount of water.

4. Use your glass-stirring rod to crush any lumps, which contain trapped sulfuric acid.

5. Stirring for a few minutes also allows for the agglomeration of colloidal particles. This permits faster, more efficient filtration.

6. Attach your filter flask to the water aspirator, immediately clamping it to avoid breaking it, add the Neoprene Filter-Vac, a 2.5 inch Büchner funnel and a piece of 5.5 cm filter paper. No trap is required since the filtrates will be discarded anyhow. (If the filtrate were to be saved, then a trap would be included to prevent a possible backup of tap water into the flask.)

7. Wet the paper, center it carefully, and turn on the water for a rapid flow. The paper should seal snugly when the funnel is pressed down if the water is running fast enough.

8. Pour about half the liquid into the funnel and press down so that the liquid filters rapidly. Swirl the beaker to loosen the crystals and pour the contents rapidly into the Büchner funnel. (If you pour slowly and carefully, most of the solid will, unfortunately, remain behind in the beaker.)

9. Residual solid may be rinsed out by using several small portions of fresh water to make a quantitative transfer. Press the filter cake with your spatula to extract the remaining liquid from the solid. A small amount of solid residue may be left in the beaker in this case.

10. Return the moist filter cake and paper to the beaker, add 50 mL of distilled water, fish out the paper filter with a glass rod, and again crush any lumps. This *trituration* removes impurities (H_2SO_4 and HNO_3) much more efficiently than merely pouring water over the crude product while filtering.

11. Collect the washed product by suction filtration. Rinse the beaker with several small portions of water to transfer as much product to the funnel as possible.

12. Place a second piece of filter paper over the filter cake and press dry with a small beaker while applying maximum suction. Pour this aqueous filtrate down the drain. Rinse the flask promptly but carefully to prevent acid holes in your clothes.

13. Weigh the product in a weighing boat, remembering to tare off the weighing boat. You should have calculated the theoretical yield (in grams) before you came to lab. Calculate the percent yield. Report the melting point range of the dried product. If dry enough, the product should melt above 65°C. Wash up for next week.

Calculation of Theoretical Yield

(To be done before lab)

Determine the limiting reagent. Calculate the moles of each reagent by dividing the grams used by the molecular weight for each reagent and by its coefficient in the balanced chemical equation.

$$\text{moles } X = \frac{\text{grams } X}{\text{molecular weight } X}$$

If pure liquids were measured by volume, calculate the grams used by multiplying the volume (mL) by the density (g/mL).

$$\text{Grams } X = (\text{Volume of } X)\,(\text{Density of } X)$$

The number of moles in solutions (e.g., HNO_3) is calculated from the molarity and volume in liters. The limiting reagent defines the maximum amount of product you can expect from a reaction, and will have the smallest number of adjusted moles (moles/coefficient).

$$\text{Moles } X = (\text{Molarity of } X)\,(\text{Liter of } X)$$

Determine the theoretical yield. Multiply the smallest number of adjusted moles from above by the molecular weight of the product and also by its coefficient in the balanced chemical equation. In the lab, the percent yield is then calculated by dividing the actual yield (in grams) by the theoretical yield and multiplying by 100.

$$\text{Percent yield} = \left(\frac{\text{actual yields}}{\text{theoretical yield}}\right)(100)$$

As an example, take the following reaction: 20 g of acetyl chloride is reacted with 100 g of ethanol to yield 10 g of ethyl acetate. First we must determine the limiting reagent. The balanced equation is:

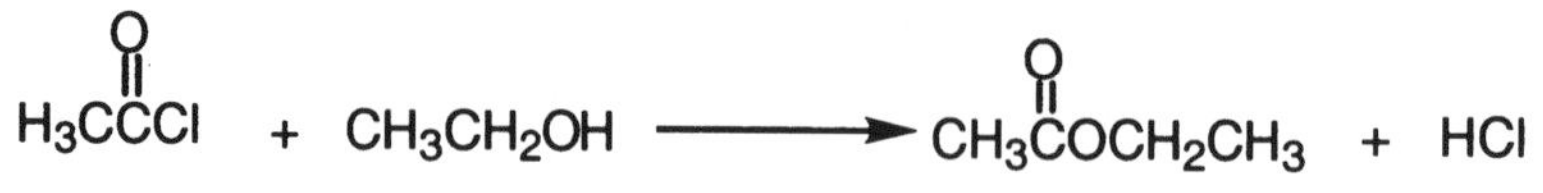

$$H_3CCCl \; + \; CH_3CH_2OH \longrightarrow CH_3COCH_2CH_3 \; + \; HCl$$

Figure 5.3 Preparation of ethyl acetate.

From this equation we know that one equivalent of both acetyl chloride and ethanol are needed to give one equivalent of ethyl acetate. Now, we must calculate the number of moles of each present, in order to determine the limiting reagent. For acetyl chloride:

$$\frac{20g}{78.5\,g/mol} = 0.25 \; mol \text{ acetyl chloride}$$

For ethanol:

$$\frac{100g}{46\ g/mol} = 2.2\ mol\ \text{ethanol}$$

Therefore, the acetyl chloride is the limiting reagent. Since one molecule of acetyl chloride leads to one molecule of ethyl acetate the theoretical yield of ethyl acetate is also 0.25 moles. This allows us to calculate the theoretical yield in grams:

$$(0.25\ mol)(88\ g/mol) = 22\ g\ \text{(theoretical yield of ethyl acetate)}$$

Since we obtained only 10 grams of product the percent yield is:

$$\frac{10g}{22g} = 45\%\ \text{yield}$$

All theoretical yield calculations are similar to the above example.

Hand in these questions at the beginning of the next laboratory period.

Questions: _____ **Name:** _____ **PID:** _____

1. If you had used toluene instead of methyl benzoate in this reaction, what nitration product(s) would have formed? Write a stepwise mechanism for the nitration reaction of toluene.

2. What does SDS stand for?

3. What are some of the physical hazards of methanol (to be used during the recrystallization of methyl 3-nitrobenzoate) as indicated in its SDS?

4. Calculate the percent yield of methyl 3-nitrobenzoate you prepared. Show all steps.

EXPERIMENT
6

Molecular Models Report

Name: _____

Partner: _____

Partner: _____

Each student must submit a report for grading.

Section No.: _____

Instructor: _____

Review of stereochemistry terms and definitions:

Constitutional isomers (Structural isomers) are molecules that have the same molecular formula, but have different connectivities.

Compounds **A** and **B** are an example of constitutional isomers

Conformational isomers (conformers) are different temporary structures of a molecule that are obtained upon rotating about a carbon–carbon single bond and/or when doing ring flips in cyclic compounds such as chair flips in cyclohexane. Also, different Newman projections of a molecule obtained from rotating about a single bond are called conformers as well.

An example of conformational isomers

37

Stereoisomers are molecules that possess the same molecular formula and the same exact structures (same connectivity), but have different spacial arrangement (different bond orientations).

A stereogenic center is a carbon that has four different atoms or four different groups of atoms attached to it.

A stereogenic center

A molecule is called **chiral** if it is not superimposable on its mirror image. A molecule is achiral (not chiral) if it has a plane of symmetry in any of its conformers.

Enantiomers are a pair of molecules that are non–superimposable mirror images of each other.

E   **F**

Compounds **E** & **F** are a pair of enantiomers.

Erythro Enantiomers are a pair of enantiomers that have the same groups on the same sides.

Erythro Enantiomers

Threo Enantiomers are a pair of enantiomers that have the same groups on the opposite sides.

Threo Enantiomers

Diastereomers are stereoisomers that are not a mirror image of one another.

G   **H**

Compounds **G** & **H** are an example of **Diastereomers**

Meso–**compounds** are molecules that have two or more stereogenic centers, but possess a plane of symmetry and therefore are achiral (Not Chiral).

A *Meso*—compound

Newman Projection

Staggered Gauche   **Staggered Anti**   **Eclipsed**

Fischer Projection

Fischer Projection

D-Glyceraldehyde   **L-Glyceraldehyde**

Assigning R (rectus, right) and S (sinister, left) Configurations:

1. Assign priorities 1 through 4 to each of the atom (groups) attached to a stereogenic center according to Cahn, Ingold, and Prelog rules. The important thing to remember about Cahn,

Ingold, and Prelog rules is: the higher the atomic number, the higher the priority and in the case of different isotopes, the one with the higher atomic masse would be assigned a higher priority. Also, one must do the comparecence one atom at a time and when there is a tie, the next set of atoms are compared until a priority is assigned.

2. Take the lowest priority (4) to the back (dashed lines), and while facing the molecule trace 1 to 2 to 3.

If the trace is clockwise, it is **R**, and if counterclockwise, it is called **S** configuration.

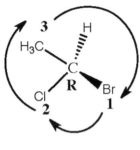

Clockwise

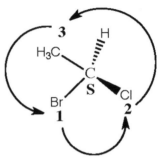

Counterclockwise

Part I. Conformations

1. Using the plastic molecular model parts, prepare a model of hexane. Use black wooden balls for carbon atoms, white or yellow for hydrogen atoms, and brown or green for bromine atoms.

    a) How many different staggered conformations exist about the C2-C3 bond of hexane?

    b) Draw the *Newman projection for the most stable conformations of* hexane looking from C2 to C3.

    c) Label each of the staggered conformers as **Anti** or **Gauche**.

    d) Does this straight chain alkane exist as a straight line?
        Explain.

e) Draw the Newman projection for the "gauche" and "anti" conformers of *n*–hexane looking from C2 to C3 bond. Which one is the most stable conformer?

2. Construct a model of *cis*–1-*sec*–butyl–3–methylcyclohexane and *tran*–1-*sec*–butyl–3–methylcyclohexane.

    a) Can you make the above molecules lie in a plane?

    b) Twist the framework molecular model (the plastic one) into the chair form. Are the various carbon bonds staggered or eclipsed?

    c) Flip the model of the cyclohexane chair conformation back and forth and slowly to observe the "chair flip." Note how the axial bonds are transformed into equatorial ones and equatorial bonds are converted to axial ones. Draw the structures resulting from a chair flip of *cis*–1-*sec*–butyl–3–methylcyclohexane and place a circle around the most stable one.

3. a) Draw the structure of the chair conformers for each of *cis*–1-methyl–4–propylcyclohexane and *tran*-1-methyl–4–propylcyclohexane and place a circle around the most stable one.

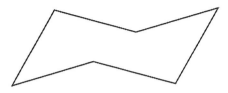

cis–1–methyl–4–propylcyclohexane

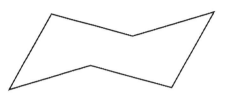

trans–1–methyl–4–propylcyclohexane

b) Repeat the above with *cis/trans* configurations of 1-methyl–3–propylcyclohexane and that of 1-methyl–2–propylcyclohexane.

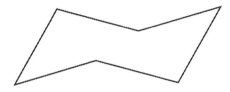

cis–1–methyl–3–propylcyclohexane

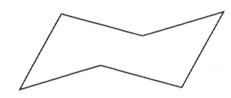

trans–1–methyl–3–propylcyclohexane

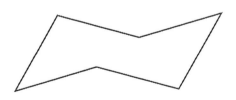

cis–1–methyl–2–propylcyclohexane

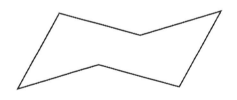

trans–1–methyl–2–propylcyclohexane

c) Draw the Newman projection for the most stable conformer of both *cis* and *trans* conformers *of* 1-methyl–3–propylcyclohexane viewing from C1–C6 & C3–C4 simultaneously.

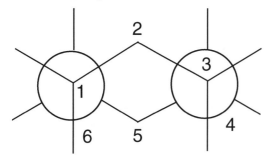

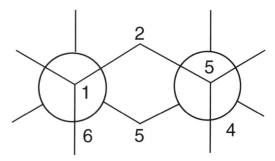

cis–1–methyl–3–propylcyclohexane   *trans*–1–methyl–3–propylcyclohexane

Part II   Stereoisomerism

4. Assemble two identical plastic models of *(R)*-2-bromo-pentane. Assume that the red ball represents bromine, yellow (or white) is hydrogen, and black is carbon. Switch any pair of atoms attached to C-2 in one of the models above. How are the two models related to one another now? (**Hint: diastereomers, enantiomers, or identical**)

5. Draw the Fischer projection formulas for *(R)*-2-bromopentane and *(S)*-2-bromopentane.

6. Make a model for each of (2R,3R)-2,3-dibromopentane and (2S,3S)-2,3-dibromopentane.

   a) What is the relationship of *(2R,3R)*-2,3-dibromopentane and *(2S,3S)*-2,3-dibromopentane?

   b) Draw the Fischer projection for both (2R,3R)-2,3-dibromopentane and (2S,3S)-2,3-dibromopentane.

7. Make a model for (3R,4S)-3,4-dibromohexane.

   a) How many stereogenic centers does this molecule have?

   b) Is this molecule chiral or achiral?

c) Are all molecules with stereogenic centers also chiral?

d) What special name is given to molecules with such structures as (3R,4S)-3,4-dibromohexane?

e) If only one of the two bromine atoms in the above structure is replaced with a chlorine atom to make (3R,4S)-3–bromo-4-chlorohexane, would the description in "d" still apply? Explain by drawing the Fischer projection for (3R,4S)-3–bromo-4-chlorohexane.

f) Draw the Fischer projection for the enantiomer of (3R,4S)-3–bromo-4-chlorohexane and properly name it.

g) Draw the Fischer projection for (2R,3S,4S)–2,3,4–trichlorooctane.

Isolation of an Active Drug

Introduction

Acetaminophen (shown below) is one of the most common over-the-counter painkillers in the market. It is the active ingredient in Tylenol and Excedrin, and many other products, although it is not the only chemical present in these medicines. There are many inert substances present as well (for example, starch is commonly used to bind the ingredients so that they may be sold in tablet form). In this lab you will isolate acetaminophen from a commercially available painkiller. This will give you additional material for discussion during the consumer product poster session.

Aspirin

Acetaminophen

Ibuprofen

Experimental Procedure

1. Use a small mortar and pestle to crush and grind one tablet completely to powder.

2. Transfer the powder to a 5-mL conical vial; be sure to thoroughly scrape the residue from the mortar.

3. Add methanol to the 2 mL mark.

4. Cap the vial (Do not use an O-ring!) and mix with an automatic mixer for 30 seconds and set aside to let the solid settle.

5. Set up a chromatography column as follows (analogous to the drying column you used previously):

   a) Insert a piece of cotton inside a Pasteur pipette and clamp it.

   b) Place your micro funnel in the Pasteur pipette.

   c) Add about 1/4 inch of sand in the Pasteur pipette.

   d) Add one inch of alumina in the Pasteur pipette.

   e) Add 1/4 inch of sand in the Pasteur pipette.

6. Place a 10-mL round bottom flask under the Pasteur pipette (the chromatography column).

7. Add 2 mL methanol to the column and allow it to pass through until the level of methanol has dropped to the top of the sand (the sand on the top).

8. Then, carefully transfer the supernatant liquid from your Tylenol–methanol mixture to the column using a Pasteur pipette (any solid transferred will clog the column and slow you down).

9. Carefully, force the liquid from the column with a pipette bulb (your lab instructor will demonstrate it).

10. Use 1 mL of methanol to rinse the insoluble ingredients in the 5-mL conical vial and add the rinse to the column.

11. Once all of the liquid has been drained from the column, add a boiling chip to the 10-mL round bottom flask containing the acetaminophen solution and distill all but 1 mL of the methanol by simple distillation.

12. Disassemble the distillation set up and remove the boiling chip with a spatula, if no solid is present scratch the inside of your vial gently with a glass stirring rod to induce crystallization.

13. Collect your product by vacuum filtration (Figure 4.1) and allow them to air dry for 5 minutes.

14. Weigh your final product and determine its melting point.

15. Calculate your percent recovery based on 500 mg acetaminophen in each Tylenol tablet.

16. Since you are working in pairs, one member of the pair should get the experimental setups prepared (such as the chromatography column) ahead of time so they are ready as needed.

Laboratory Safety

Although you are isolating a commercial drug, your isolated product will not have passed appropriate testing for purity and safety and should NOT be ingested. Methanol (wood alcohol) is known to cause blindness and other problems (including death) when ingested, handle with care and as always keep your goggles on. The distillation should be conducted under the hood.

Questions: _____ Name: _____ PID: _____

1. Product

    a) What is the final mass of acetaminophen crystals?

    b) What is the percent recovery of your product?

    c) If your percent recovery is more than 100%, propose reasons for it. If not, list possible sources for the loss of product.

d) What is the melting point of your product?

The literature melting of acetaminophen is 170°C.

What can you say about the purity of your product?

EXPERIMENT 8

Azo Dye Synthesis, Orange II

Background and Discussion

Azo dyes are among the most common classes of dyes used commercially. They are relatively simple to make, and small modifications in the functional groups of the reactants can result in very different dye colors. This permits a range of colors to be obtained with few modifications to the basic reaction. Orange II (Figure 8.1, **VI**) is a typical azo dye, and is obtained from relatively inexpensive starting materials. The IUPAC name for Orange II is 4-[(2-hydroxy-1-naphthalenyl) azo]benzene sulfonic acid monosodium salt. It is also known as C.I. Acid Orange 70, C.I. 15510, Tropaeolin 000 no. 2, and Betanaphthol orange. Such multi-name chaos is common with both dyes and pharmaceuticals.

The final product **VI** exists in three possible forms, depending on the pH. In strong aqueous acid, both acid sites are protonated, and the molecule is magenta. The stable monoanion isolated in this experiment is bright orange. In strongly basic solutions, both of the acidic protons in sulfonic acid and phenol are removed by the base. This very soluble dianion is reddish brown. Measure the various acidic and basic reagents carefully, or you will get the wrong product.

This synthesis of VI (Figure 8.1) involves a diazonium coupling of sulfanilic acid (**I**) with the sodium salt of 2-naphthol. The oxidation of the $-NH_2$ group to the diazonium ion ($-N_2^+$) must be performed in acid, even though sulfanilic acid is not soluble in acid. To overcome this problem, the sulfanilic acid is dissolved in base and precipitated as a fine suspension in the acid solution. The sodium nitrite is acidified at the same time and immediately reacts with the amine.

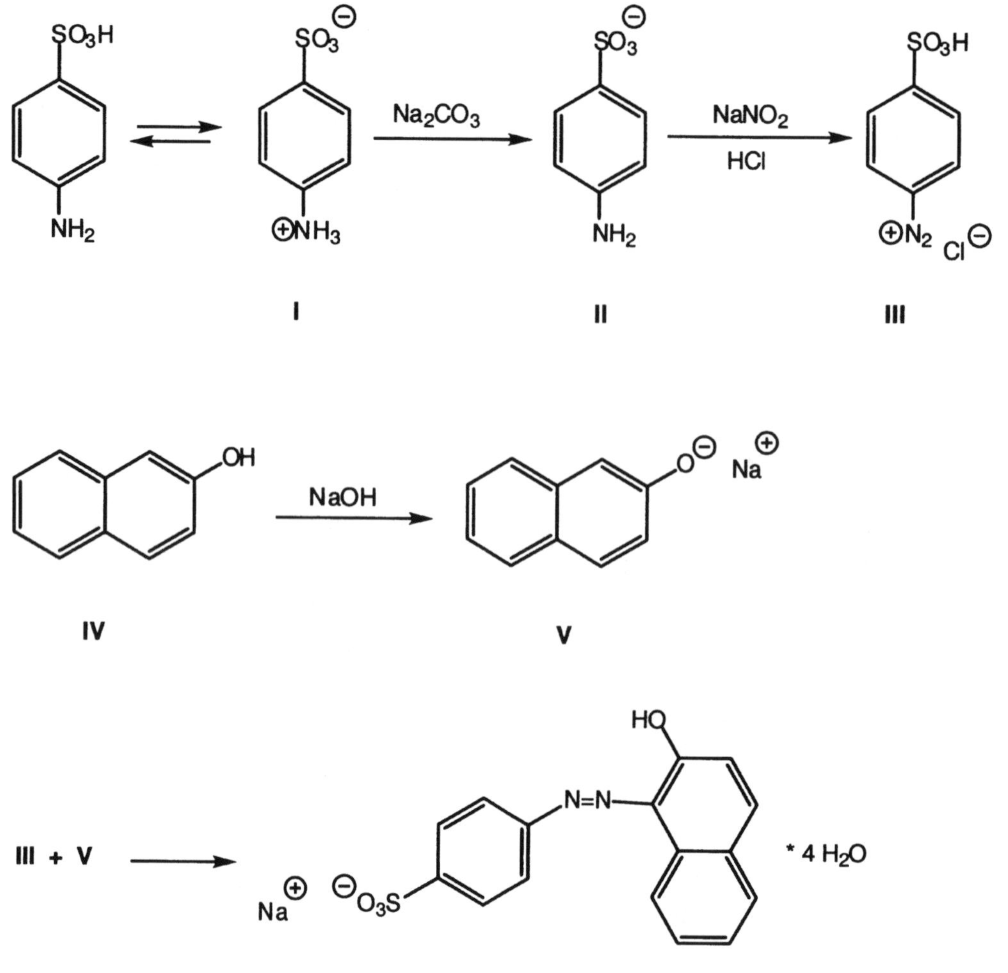

VI, Orange II

Figure 8.1 Synthesis of Orange II

The mechanism of diazotization is known to be a radical mechanism but the overall process is not completely understood. A cold acidic solution of $NaNO_2$ generates the relatively unstable nitrous acid. This species performs the oxidation of the arylamine to the diazonium salt. The solution must always be kept cold (0°C) to prevent the water present from acting as a nucleophile, producing phenols.

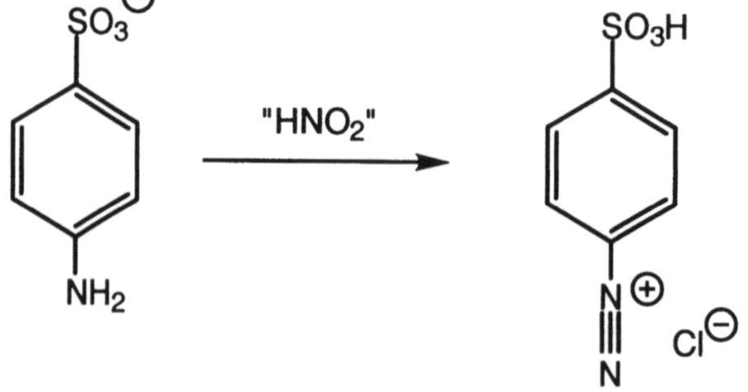

Figure 8.2 Diazotization of sulfanilic acid.

As shown in Figure 8.3, the actual mechanism of azo-coupling involves nucleophilic attack by the *ortho* site of the phenol on the diazonium salt. It is actually the free phenol that attacks not the phenoxide salt. Although the phenoxide salt must be made in order to dissolve the naphthol in aqueous solution, the equilibrium between the salt and free phenol is such that sufficient free phenol is available to obtain a good yield of the coupled product.

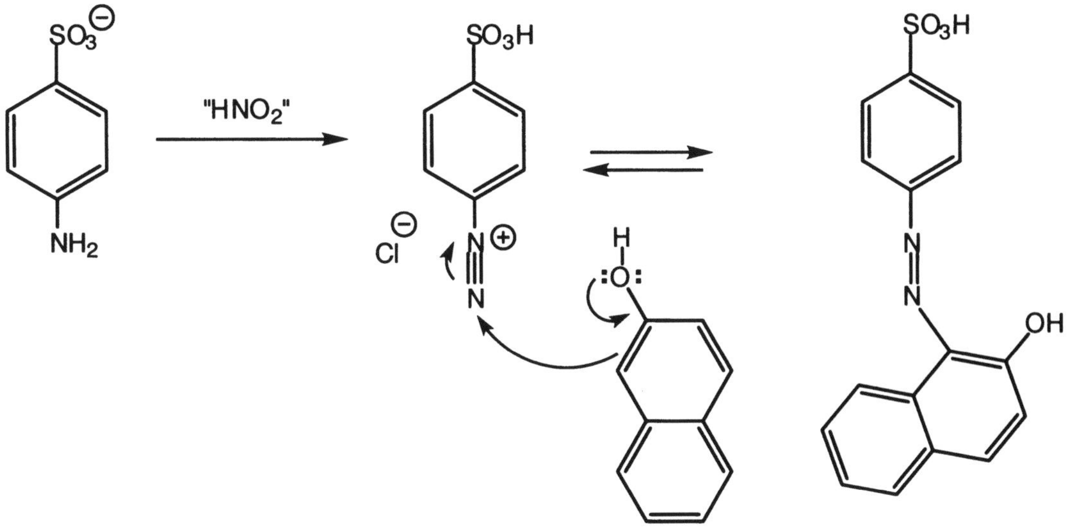

Figure 8.3 Mechanism of azo–coupling.

Diazonium salts are unstable, even when conjugated with aromatic systems. Control of temperature is critical once the sulfanilic acid is diazotized. Keep the diazonium salt below 10°C and add the sodium 2-naphthoxide solution as soon as possible. If the temperature rises above 10°C, or if the diazonium salt is stored for any length of time, considerable hydrolysis to the phenol and nitrogen gas will occur. Thermoprobes should be used to monitor temperatures. In this experiment, it is sufficient to keep a couple of pieces of ice inside the reaction flask at all times. Finally, diazonium decomposition is catalyzed by some metal ions. Use a glass rod, not a metal spatula, to stir the various solutions.

Experimental Procedure

Calculate the theoretical yield before the lab.

Caution: Orange II dyes clothing and other organic materials a permanent, unique orange. As a measure of your technique, points may be subtracted for orange hands, etc.

I. Diazotization of Sulfanilic Acid

1. Dissolve 0.4 g of anhydrous sodium carbonate in 7.5 mL of H_2O. Use a 25-mL Erlenmeyer flask.

2. Add 0.45 g of sulfanilic acid to the solution, and heat over a steam bath until it dissolves.

3. Cool the filtrate to **room temperature**, add 0.2 g of sodium nitrite and stir until it also dissolves.

4. Pour this pale yellow solution slowly, using a stirring rod, into a 125 mL Erlenmeyer flask containing 2 mL of H_2O, 1 mL of concentrated HCl (12 M) and about 3 g of ice or enough to just cover the bottom of the flask.

5. In a short time, the diazonium salt of sulfanilic acid **may** separate as a finely divided white precipitate.

6. Keep this suspension cooled in an ice bath until it is to be used. More ice may be added to the flask if needed; keep a few pieces in the flask at all times.

II. Orange II

1. Combine 0.38 g of 2-naphthol and 1.6 mL of 10% (2.5 M) sodium hydroxide in a 10 mL pear-shape flask.

2. Warm over steam to hasten the dissolving (1–2 min.); then cool briefly in an ice bath (30 seconds). If the sodium salt of 2-naphthol crystallizes, reheat gently to dissolve any solids.

3. Using a dropper, add this solution to the cold suspension of diazotized sulfanilic acid in the 125-mL flask.

4. Stir thoroughly by swirling. In a few minutes, the Orange II will begin to crystallize.

5. Keep the mixture in an ice bath for another 15 minutes to permit the coupling reaction to approach completion. Allow the remaining ice in the mixture to melt.

6. Collect the crude Orange II by suction filtration using a **4-inch** Büchner funnel. Do not wash this salt with additional water.

7. Pump dry for at least 5 minutes or until filter cake is as dry as possible. If much of the solid remained in the 125-mL flask, pour about 1/2 of the filtrate back into the flask.

8. Filter the suspension in the 4-inch Büchner funnel. Be sure to use a new filter paper. Dry this filter cake as well.

9. Weigh the dried dye in a weighing boat. The percent yield is to be calculated. Show all of your calculations. Show your product to your instructor and be sure they have recorded this fact, before disposing of the dye in the solid hazardous waste container.

Questions: _____ **Name:** _____ **PID:** _____

1. What is the purpose of adding sodium carbonate in the diazotization of sulfanilic acid?

2. Would you need to use sodium carbonate for the diazotization of *m*–nitroaniline? Explain.

3. Calculate the theoretical yield for Orange II.
   (MW of Orange II = 422 amu, show your work).

EXPERIMENT 9

Synthesis of Aspirin/ Preparation of Soap

Introduction

Your task during this lab is similar to that during the azo dye experiment, you will synthesize organic compounds. You will use reactions between nucleophiles and electrophiles to prepare aspirin and soap. Aspirin is an antipyretic (fever-reducer) and analgesic (painkiller) that is structurally similar to a natural painkiller found in willowbark and was in use for hundreds of years before aspirin was available. Aspirin is one of the oldest over–the–counter drugs. Its worldwide production exceeds 50 billion tablets annually.

Aspirin is synthesized by reacting salicylic acid with acetic anhydride in an acidic medium (Figure 9.1).

Salicylic Acid	Acetic Anhydride	Acetylsalicylic Acid (Aspirin)

Figure 9.1 Preparation of Aspirin.

Natural soaps are synthesized from animal fat such as lard by saponification (in Latin sapo means soap), which is the reaction of triglycerides with sodium hydroxide (Figure 9.2). Soaps have been used for centuries as cleansing agents. Soaps are simply the sodium salts of fatty acids. The sodium salt keeps the soap soluble in water and the hydrocarbon part of it dissolves "dirt." Fatty acids have the following general properties:

1. All naturally occurring fatty acids have an even number of carbons (12–20 carbons).

2. In most unsaturated fatty acid, *cis* double bonds predominate.

3. Saturated fatty acids have higher melting points than their respective unsaturated ones.

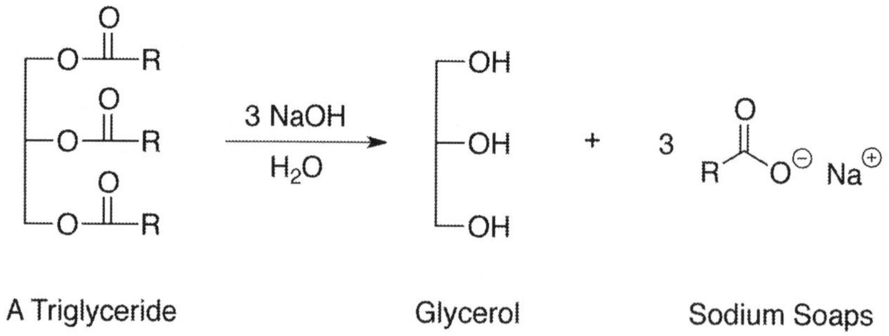

Figure 9.2 Preparation of sodium soaps.

Experimental Procedure

I. Aspirin

1. Set up a hot water bath at about 90°C.

2. Add 0.105 g of salicylic acid, 0.240 mL of acetic anhydride, 1 drop concentrated phosphoric acid, and a magnetic stir-bar to a 10–mL round-bottom flask.

3. Attach an air condenser and clamp the apparatus so that the bottom of the round-bottom flask is in the hot water bath.

4. Heat the mixture for 4–5 minutes after the salicylic acid has dissolved completely.

5. Remove the flask from the water bath and add 0.5-mL cold water slowly through the top of the air condenser. This will hydrolyze any excess acetic anhydride.

6. Remove the air condenser and retrieve your magnetic stir-bar with forceps.

7. Rinse the residue from the stir-bar into your round-bottom with 1-mL water; add another 1 mL of water and chill in an ice bath. *You should be careful when washing your magnetic stir bar over the sink. It is easy to lose them down the drain.*

8. Collect the product by vacuum filtration, rinsing with 0.5 mL cold water several times.

9. Weigh product after air–drying and determine its melting point.

II. Soap

1. Make your soap from 1 mL of Crisco oil, 2 mL of 50% sodium hydroxide solution, a couple of boiling chips, and 2 mL of ethanol in a 50 mL Erlenmeyer flask.

2. Reflux over a steam bath, swirling occasionally to knock material back down into the flask.

3. When you have a homogeneous solution the saponification is complete.

4. Pour the mixture quickly into 25 mL of a premixed saturated salt solution. (The Erlenmeyer flask will be hot so hold it with a clamp while pouring.)

5. Stir several minutes and collect the soap by vacuum filtration. Rinse with two 10 mL portions of cold water and air dry.

Laboratory Safety

Sodium hydroxide, acetic anhydride, and phosphoric acid are corrosive, wear gloves while handling. The aspirin you will make is not free of impurities. Do **NOT** ingest it or any other chemicals you work with. Also, the soap you have made is not to be used for cleaning. You must turn in all your products to your TA otherwise; you will not get credit for the experiment.

Questions: _____ **Name:** _____ **PID:** _____

1. Aspirin

    a. Draw the structure of acetic anhydride and identify any polarized bonds.

    b. Draw the structure of salicylic acid and identify all nucleophilic sites.

2. What is the purpose of adding phosphoric acid in the synthesis of Aspirin?

Questions: _____ Name: _____ PID: _____

3. Soap

   a. Assume that Crisco is a single triacylglyceride (it is actually a mixture of triacylglycerides) in which all acyl groups are palmitate (palmitic acid = $CH_3(CH_2)_{14}CO_2H$). Draw the structure of this triacylglyceride.

   b. Show the products that are generated when Crisco and sodium hydroxide react. Use the same assumption as in question a.

   c. What is the purpose of adding ethanol in the preparation of soap?

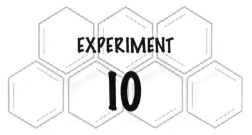

EXPERIMENT 10

Identification of an Unknown by Derivatization

Background and Discussion

Before the development of advanced spectroscopic techniques the preparation of a solid derivative of a compound was the best method of identifying an unknown compound. First, using various chemical tests, the functional group class of the unknown compound was identified. This information, along with any data on the physical properties of the unknown, would help to decide which derivative to prepare. After the derivative was made, its melting point could be compared to those of known derivatives of that type and identification is made. Occasionally more than one derivative was needed to effect positive identification.

One of the most valuable derivatives of aldehydes and ketones are the hydrazones. They are formed by reaction of the carbonyl compound with a hydrazine. For instance, 2,4-dinitrophenylhydrazine reacts quickly and quantitatively with aldehydes and ketones (Figure 10.1). This is of great utility to derivatizations as often only a tiny amount of the unknown is available. The most commonly used hydrazine for derivatization is 2,4-dinitrophenyl hydrazine since it gives a solid product in virtually all cases.

Figure 10.1 Formation of 2,4-dinitrophenylhydrazone derivative.

Aldehydes and ketones also react with semicabazide to form the semicarbazone derivatives (Figure 10.2). Note that the reaction occurs exclusively at one end of the semicarbazide. The amino group (NH$_2$) that is next to the carbonyl group does not participate in this reaction. Why?

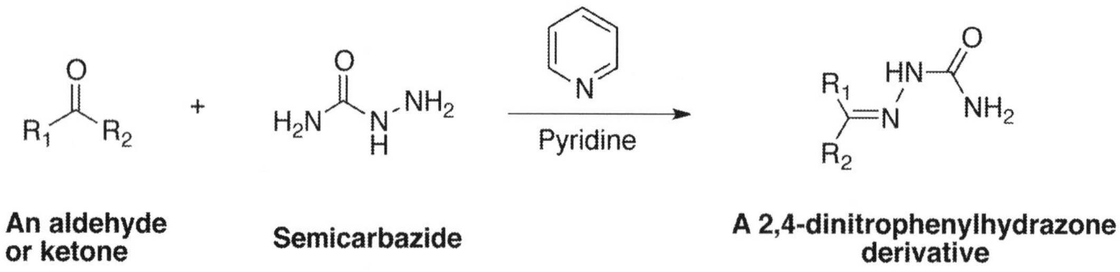

An aldehyde or ketone + **Semicarbazide** → (Pyridine) **A 2,4-dinitrophenylhydrazone derivative**

Figure 10.2   Formation of a semicarbazone derivative

Experimental Procedure

The identification of substances is a common problem encountered in the final stages of structure determination of a compound of unknown structure. We have tried to set up this experiment as a realistic experience in identification subject to the limitation of time, materials, and pedagogy. Also note, cleaning acetone is a ketone and will react with both reagents, do not clean the flasks with acetone beforehand!

I. Synthesis of a 2,4-Dinitrophenylhydrazones Derivative

CAUTION: 2,4-DNP causes permanent stains.

1. Add 0.3 mL (about 10 drops of the liquid) of your unknown into a 25–mL Erlenmeyer flask.

2. Pipette 3 mL of the 2,4-dinitrophenylhydrazine reagent solution directly into the Erlenmeyer flask, using the supplied pipette. DO NOT use a graduated cylinder, as the 2,4-dinitrophenylhydrazine reagent solution is messy.

3. Stopper the flask with a well fitting cork, swirl briskly to mix and let it stand undisturbed for 10 minutes or until sufficient solid has precipitated out.

4. If crystals have not yet appeared after 10 minutes, remove the cork, and heat gently over steam for fifteen minutes.

5. Allow the warm solution to cool to room temperature and scratch the inside surface of the Erlenmeyer flask using a glass stirring rod to induce crystallization.

6. Collect the crude product by vacuum filtration on a Hirsch funnel.

7. Pour this filtrate into the hazardous waste container.

8. Triturate the crude yellow or orange product as follows:

    a) Transfer the impure hydrazone derivative to a 50 mL beaker.

    b) Add 10 mL of 2 M hydrochloric acid to the 50 mL beaker.

    c) Crush the solids using a glass-stirring rod in order to remove any unreacted red 2,4-dinitrophenylhydrazine (mp = 200°C).

    d) Filter the crude product by suction filtration and rinse it with 20 mL of water and finally with a little ice–cold ethanol.

    e) These rinsings may go down the drain.

9. Take the melting point of the solid DNP derivative.

II. Synthesis of a Semicarbazone Derivative

1. Prepare a semicarbazone by mixing 3.0 mL of the semicarbazide reagent and 0.3 mL (about 10 drops) of your unknown sample.

2. Warm your solution on a steam bath until crystals begin to form.

3. Cool and collect the product by vacuum filtration with a Hirsch funnel.

4. The filtrate is to be poured into the hazardous waste container.

5. Take the melting point of the solid semicarbazide derivative.

Remember in this experiment your grade only depends on the purity of the derivatives and not the amount of it. Therefore, be very careful in measuring the melting points of your two derivatives. You may compare your melting point values with the literature values to help you in the identification of your unknown sample. The literature values of selected 2,4–dinitrophenylhydrazone (2,4-DNP) and semicarbazone derivatives of a series of aldehydes and ketones are provided in Table 10.1.

Compound	2,4-DNP Derivative (°C)	Semicarbazone (°C)	Boiling Point (°C)
Butanal	122	104	75
Heptanal	108	109	156
Furfural	229	202	162
Benzaldehyde	237	222	179
Cinnamaldehyde	255	215	252
Octanal	106	101	171
2-Butanone	117	146	80
2-Pentanone	144	110	102
3-Pentanone	156	139	102
Cyclopentanone	142	205	131
Cyclohexanone	162	166	155
Acetophenone	250	198	202
Propiophenone	191	174	218
4-Methylacetophenone (methyl p-tolyl ketone)	260	205	226
3-Methyl-2-butanone	120	114	94
4-Methyl-2-pentanone (isobutyl methyl ketone)	95	135	119
3-Hexanone	130	113	125
2-Hexanone	110	122	129
2,4-Dimethyl-3-pentanone	107	160	125
3,3-Dimethyl-2-butanone (pinacolone)	125	158	106
3-Methyl-2-pentanone	71	95	118
Isobutyrophenone	163	181	222
1-Phenyl-2-propanone (benzyl methyl ketone)	159	210	216
Butyrophenone	190	191	230

Table 10.1   Solid derivatives (*2,4–Dinitrophenylhydrazone, 2,4-DNP and Semicarbazone*) of some aldehydes and ketones.

Questions: _____ Name: _____ PID: _____

1. What is the melting point of 2,4-DNP derivative of your unknown sample?

2. What is the melting point of semicarbazone derivative of your unknown sample?

3. What is the most likely identity of your unknown?

   (Use Table 10.1)

4. In the reaction of semicarbazide with aldehydes and ketones the amino group (NH_2) that is next to the carbonyl group does not participate in the reaction. Explain the reason.

EXPERIMENT 11

Structure and Reactivity of Carbohydrates

Introduction

Carbohydrates are polyhydroxy aldehydes and ketones or molecules that are hydrolyzed to produce polyhydroxy aldehydes and ketones. They are extremely important biological molecules and the most abundant in nature. Among their many functions, they are energy sources (glucose, starch), supportive structure of plants (cellulose), and they provide the backbone structure of DNA (D–2–deoxyribose) and RNA (D–ribose).

The simpler member of carbohydrates are called saccharides (Latin: saccharum = sugar). Monosaccharides such as glucose and fructose (Figure 11.1) are the simplest Carbohydrates that cannot be hydrolyzed into smaller sugars. Disaccharide is hydrolyzed to yield two molecules of Monosaccharides. Trisaccharide affords three molecules of monosaccharides when hydrolyzed. Molecules that are made of 4 to 10 units of monosaccharides are called oligosaccharides and molecules that are made of 11 or more monosaccharides are called polysaccharides. Monosaccharides are classified based on the number of carbon atoms in the molecule and the type of the functional groups. For instance, glyceraldehyde is an aldotriose and fructose is a ketohexose (Figure 11.1).

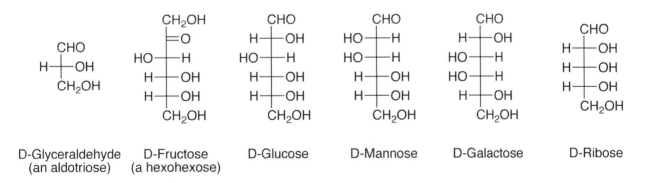

Figure 11.1 Fisher projection of selected monosacharides.

Carbohydrates that are oxidized with mild oxidizing reagents such as Benedict or Tollens reagents are called reducing sugars. Tollens reagent is made from silver oxide or silver nitrate and ammonium hydroxide. Benedict solution is an aqueous solution of copper sulfate, sodium citrate, and sodium carbonate. In this laboratory you will explore the relationship between the structure of various sugars and their reactivity toward oxidizing agents. Review Chapter 16 in your text and pay special attention to section 16.10, oxidation of monosaccharide while preparing for this laboratory.

Experimental Procedure

1. Your instructor will assign four compounds to you and your partner from the list below.

2. Build a model of one of these compounds (as directed by your TA) and use it to answer question 1.

3. For each of the four carbohydrates, mix 5 mL of Benedict's solution with 10 drops of the sugar solution in a test tube.

4. Place 5 mL of Benedict's solution in a test tube with 10 drops of water as a control.

5. Heat the five test tubes in boiling water for 5 minutes and note any color changes. Dispose of all solutions in the hazardous waste container.

    Compounds: fructose, galactose, glucose, lactose, mannitol, mannose, starch, sorbitol, sucrose

Safety/Hazards

Point the opening of the test tubes in boiling water toward the hood (away from you and other people).

Questions: _____ Name: _____ PID: _____

1. **Structure**

   a. Draw the Fisher and Haworth projection for your sugar.

   b. Based on your knowledge of reactivity, is your sugar likely to be a reducing sugar?

2. **Observations**

   a. What was the initial color of Benedict's solution?

   b. What was the final color of your control sample?

   c. What was the final color of your carbohydrate sample after addition of Benedict's solution?

   d. Does the experimental evidence indicate that your carbohydrate is a reducing sugar?

EXPERIMENT 12

Consumer Product Project

Task Description: Each student will make a 5-minute oral presentation explaining their poster on the structure of a pre-approved organic molecule present on the label of a consumer product (food label, chemical label). The molecule needs to contain at least 5 carbons and must be approved by your TA. The presentation should include the original product label, the chemical structure, and chemical formula of the chosen molecule. You should also include information about its origin, chemical properties, and spectroscopic characterization of the compound. The students should clearly identify the different functional groups present on the molecule and should be able to answer questions regarding their molecule (both physical and chemical properties). The poster will be evaluated based on its organization and information content. The artistic value of the poster will also be evaluated based on both creativity and clarity. The literature references used for the project should be clearly presented as well.

	Excellent	**Competent**	**Needs Work**
Oral Presentation:			
Presentation Skills 6 points	Presenter speaks clearly and loudly to be heard, uses eye contact and body language to engage the audience	Presenter speaks clearly and loudly to be heard but sometimes tends to drone, occasionally uses eye contact and body language to engage the audience	The presenter cannot be heard and/or understood. Presenter is monotonous. No eye contact, no engaging audience
Questions/ Knowledge 8 points	Presenter can follow the questions and provides complete answers.	Presenter does not know all answers but attempts to answer the questions. Does not provide complete answers.	Presenter cannot understand questions / Does not know the answers / Does not attempt to answer.
Poster:			
Information/ Organization 8 points	The poster contains the product label and information about the origin of the compound, molecular formula, physical and chemical properties. The information is organized in a logical, easy to follow manner.	The poster contains the product label and most information about the origin of the compound, molecular formula, physical and chemical properties but one or two major components are missing. The information is organized in a logical, easy to follow manner.	Major parts of necessary information are missing (either label, origin, formula, or properties). The way the information is organized is hard to follow.
References 3 points	References for all the presented information, including spectra, food label, and properties are clearly presented either at the end of the poster or directly under the presented information.	Most references are present but some of them are missing or not clearly presented.	There are no references or if there are any they are incomplete and poorly presented.
Creativity/Clarity 5 points	The poster is presented both clearly and creatively. Use of color and good quality pictures.	The poster is presentable (good quality pictures) but lacks attention to details. Minimum use of color.	The poster is sloppy and looks like it was put together in a few minutes. No organization, poor quality pictures.

Scoring: 25–30: Excellent; 15–25: Competent; 0–15: Needs Work

Grignard Reaction: Preparation of Triphenylmethanol

Background and Discussion

The Grignard reaction was one of the first organometallic reactions discovered and is still one of the most useful synthetically. By reacting an organohalide (usually a bromide) with magnesium in ethereal solvent, carbon becomes a nucleophile—and the starting point for many efficient syntheses. Grignard reagents are the starting points for many syntheses of alkanes, primary, secondary, and tertiary alcohols, alkenes, and carboxylic acids.

The formation of Grignard reagents are extremely sensitive to moisture, therefore it is imperative that all apparatus and glassware used for their preparation be as dry as possible. Phenyl magnesium bromide is one of the easier Grignard reagents to prepare. As bromobenzene is relatively inexpensive phenyl magnesium bromide may be used economically in excess. Also, competing coupling reactions, to form biphenyl are not a major concern.

Triphenylmethanol is synthesized by reacting phenyl magnesium bromide with an ester of benzoic acid. The particular ester (methyl or ethyl esters being the most common ones) does not affect the final product, as the alcohol group is lost during the reaction. Since esters consume two equivalents of the Grignard reagent, the stoichiometry of the reaction would be:

Figure 13.1 The Stoichiometry of a Grignard reaction.

The initial reaction, between the magnesium and the alkyl halide to form the Grignard reagent, takes place via a radical mechanism. The presence of free radicals leads to the generation of biphenyl as a byproduct (Figure 13.3). The Grignard reagent reacts with remaining unreacted alkyl halide to give the dimer, biphenyl. Byproduct formation is increased by an increase in concentration of the starting alkyl halide solution.

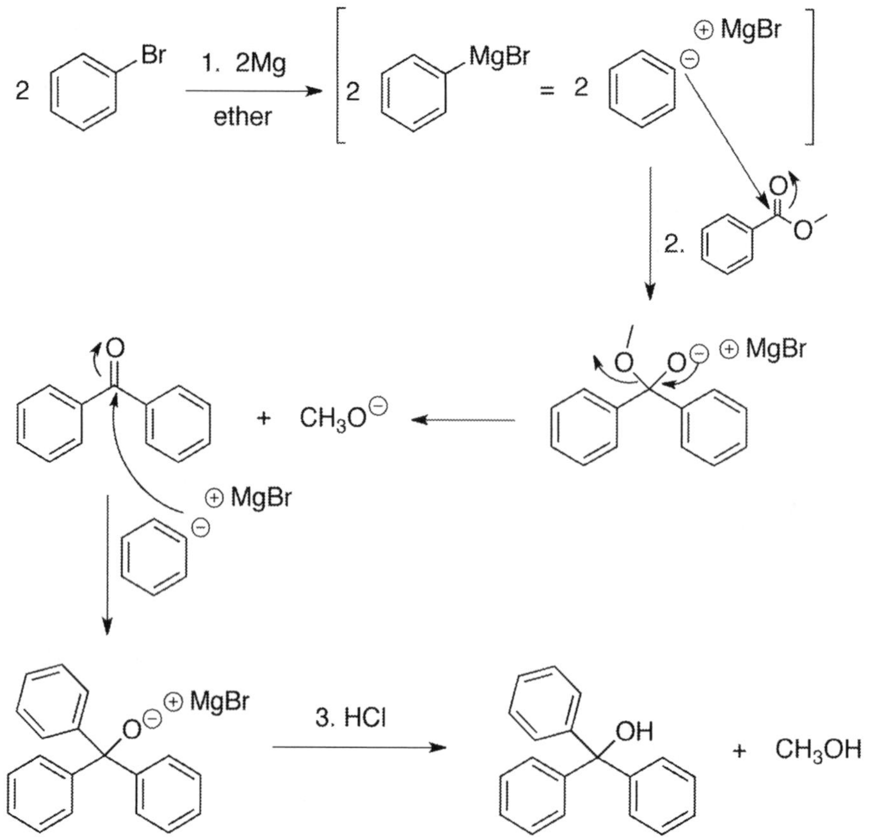

Figure 13.2   The Grignard synthesis of triphenylmethanol.

The second step in the Grignard reaction is much simpler mechanistically (Figure 13.2). The electropositive magnesium adjacent to the carbon causes the carbon to behave as a nucleophile. The Grignard nucleophile attacks the ester carbonyl to form the intermediate **I**. Loss of the methoxide ion (**II**) generates an intermediate ketone, **III** that is generally not isolable. Instead, a second Grignard nucleophile attacks the newly formed ketone carbonyl yielding the final alkoxide, **IV**. The free alcohol is generated after the alkoxide, **IV** is protonated in the acidic workup, to give the final product, triphenylmethanol, **V**.

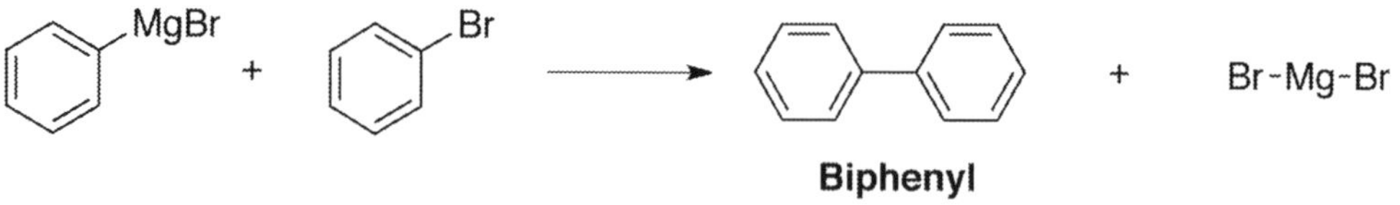

Figure 13.3   Biphenyl formation as a byproduct in a Grignard reaction.

Recrystallization of the triphenylmethanol is necessary to remove the byproduct biphenyl that forms during the reaction.

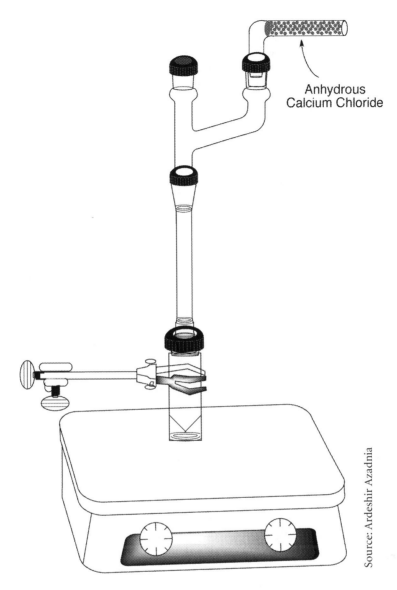

Anhydrous
Calcium Chloride

Source: Ardeshir Azadnia

Figure 13.4   Grignard experimental setup.

Caution

Preparation of the Grignard reagent, its addition to an ester, and acid workup to yield the final product, are highly exothermic reactions. Therefore, one should exercise caution when doing this experiment in larger scales. For instance, mix the reagents slowly and be prepared with a water-ice bath to moderate any over-exuberant reaction. Ether fumes will ignite if they touch any surface over 200°C (auto ignition temperature). No flames should be permitted anywhere in the lab when working with ether. Measure out the required amounts of ether as needed in the hood. Do not leave open beakers or bottles of ether on your bench top.

Experimental Procedure

All glassware used in a Grignard reaction must be scrupulously dried. *Dry the following glassware in an oven at 110°C for at least 20 minutes: drying tube, Claisen Adapter, 8 mL conical vial, 5 mL conical vial, air condenser, glass stirring rod, and a magnetic spinvane.*

Week 1

1. Assemble the oven-dried apparatus as shown in Figure 13.4.

2. Add 53 mg (2.2 mmol) of "shiny" magnesium turnings (Mg, Grignard grade).

3. Remove the air condenser equipped with a drying tube and quickly add the magnesium, 2.0 mL of anhydrous ether, and 260 μL (2.5 mmol) bromobenzene to a clean and dry 8–mL conical vial.

4. Support the bottom of the vial on a cork stopper and press firmly on the Mg–turnings with a clean, dry glass-stirring rod repeatedly to expose fresh metal to induce the reaction. When the reaction starts, the solution will turn cloudy then amber and boil spontaneously. Check with your instructor if you cannot get your reaction to start.

5. Add a spinvane to the reaction vial, replace the air condenser assembly, and tighten the cap seal.

6. Adjust the reaction vial on a hotplate stirrer and begin rapid stirring. Most of the magnesium will be gone and the solution will take on a light amber color after 5 minutes.

7. Heat slowly to reflux (hotplate setting approximately at 3–4) for an additional 5 minutes. Then cool the reaction mixture via an ice bath for 2 minutes.

8. Dissolve 0.125 mL of methylbenzoate (density = 1.09 g/mL) in 1.0 mL anhydrous ether in a 5 mL conical vial.

9. Draw the methyl benzoate solution into a clean, dry syringe.

10. Place the syringe containing methyl benzoate solution in the septum of the cap of the Claisen adapter and add the solution dropwise over 1–2 minutes.

11. Vigorous stirring of the reaction vial contents is essential.

12. Stir at room temperature for 5 minutes and then warm to reflux for an additional 5 minutes.

13. Cool the reaction vessel to room temperature and add 1 mL of dilute HCl. All solids should dissolve; if not, add 0.5 mL more dilute HCl.

14. Stir the reaction mixture for 2–3 minutes. This is a good stopping point.

15. Cap the 8-mL conical vial and place it inside of a 100 mL beaker and carefully secure the beaker in your locker until next week.

Week 2

1. Add 3 mL of ether and 1 mL of water to the 8-mL conical vial and mix it well for a few minutes.

2. Allow the water layer (in bottom) and the ether layer (on top) to fully separate.

3. Insert a small piece of cotton inside the tip of a short stem pipette using a long piece of a stainless steel wire.

4. Remove the spinvane and pipette the aqueous layer into a 4–inch test tube (ether is less dense than water).

5. Wash the ether (still in the reaction vial) with two 1 mL aliquots of water and each time discard the water layer (in bottom).

6. Prepare a short column of magnesium sulfate or sodium sulfate (Figure 4.3) to affect drying of the ether layer as follows:

   a) place a wad of cotton in a Pasteur pipette

   b) add 0.5 cm of sand

   c) add 2 cm of Na_2SO_4

   d) add 0.5 cm of sand.

7. Clamp the Pasteur pipette upright and pass the ether layer through the drying agent into a 10–mL round bottom flask.

8. Rinse the reaction vial with two 1–mL aliquots of ether and pass these through the pipette containing Na_2SO_4 in order to make the transfer quantitative.

9. Add a boiling chip to the ether solution and remove ether by simple distillation.

10. When the ether is gone, slowly add 2–3 mL of hexanes to the round bottom flask and allow the mixture to cool gradually to room temperature. It is fortunate that biphenyl (the byproduct) dissolves in hexanes well and triphenylmethanol does not at all. Therefore, the desired product may be cleaned up without recrystallization. Solids should appear before cooling in an ice bath.

11. Filter the solids using your small Hirsch funnel.

12. Weigh the dried crystals and take a melting point and decide for yourself whether you should recrystallize your triphenylmethanol from hot ethanol or not. Remember going through a recrystallization process would reduce your yield.

13. Also, when recrystallizing from ethanol (if you decide to do it), dissolve the crude triphenylmethanol in minimum amount of hot ethanol and allow the solution to cool to room temperature slowly.

14. Then, cool and collect the crystals as before.

15. Pour all of the filtrate into the hazardous waste container.

16. Spread the product on an 8 1/2 × 11 sheet of paper to dry.

17. Determine the melting point (and range) of your purified product. If the range is greater than 3°C, either the sample is impure or wet, or the melting point was improperly done.

18. Correct any flaws, and repeat the melting point.

19. Place the dry product in a properly labeled ziplock bag and turn it in to your TA with your lab report.

LABORATORY EXPERIMENT 13
Grignard Reaction: Preparation of Triphenylmethanol Report

Questions: _____ Name: _____ PID: _____

1. **(1 Point)** What is the theoretical yield and percent yield of your triphenylmethanol in this experiment?

2. **(1 Point)** Why is it essential to use a drying tube when performing a Grignard reaction?

3. **(1 Point)** Can you think of a more efficient way of keeping the moisture out of the reaction vessel, than using a drying tube?

Check Out Instructions

1. Complete "Lab Instructor Evaluation" forms and place it anonymously in the brown envelope.

2. Thoroughly clean all dirty glassware, trying soap and hot water first. Acetone, however, may be needed to remove some organic compounds, be sure it goes into the hazardous waste container, not down the drain.

3. Place all of the contents of your drawers on the bench and arrange them according to the check-in sheet.

4. Make sure you have all items on the list. If you are missing any equipment, replace it from the second floor stockroom by presenting them with your student ID.

5. Once you are sure that you have all the equipment, write your name on the board and wait for your TA to check you out of lab.

6. Do not leave until your TA has signed your check-in sheet. When the locker has been properly checked out, your instructor will close and lock it and also sign the inventory check–out form.

7. You must be checked out of the laboratory by your instructor at the end of the term or you will be charged $35.00 for failure to check out, in addition to the price of any missing items. This is true even for people who drop the course early. Even if you only went to lab one time, you still need to check out of the lab. You are required to have your MSU Student ID in order to be checked out of your laboratory locker.

Name: Date:

Experiment:

Name: Date:

Experiment:

Name: _____   Date: _____

Experiment: _____

Name: Date:

Experiment:

Name: _____   Date: _____

Experiment: _____

Name: _____   Date: _____

Experiment: _____

Name: _____   Date: _____

Experiment: _____

Name: _____   Date: _____

Experiment: _____

Name: _____   Date: _____

Experiment: _____

Name: _____ Date: _____

Experiment: _____

Name: _____   Date: _____

Experiment: _____

Name: _____   Date: _____

Experiment: _____

Name: _____   Date: _____

Experiment: _____

Name: _____   Date: _____

Experiment: _____

Name: _____   Date: _____

Experiment: _____

Name: _____ Date: _____

Experiment: _____

Name: _____   Date: _____

Experiment: _____

Name: Date:

Experiment:

Name: _____   Date: _____

Experiment: _____

Name: _____ Date: _____

Experiment: _____

Name: _____ Date: _____

Experiment: _____

Name: _____ Date: _____

Experiment: _____

Name: _____   Date: _____

Experiment: _____

Name: Date:

Experiment:

Name:

Date:

Experiment:

Name: _____ Date: _____

Experiment: _____

Name: _____   Date: _____

Experiment: _____

Name: _____   Date: _____

Experiment: _____

Name: _____ Date: _____

Experiment: _____

Name: _____ Date: _____

Experiment: _____

Name: _____   Date: _____

Experiment: _____

Name: _____   Date: _____

Experiment: _____

Name: _____   Date: _____

Experiment: _____

Name: _____   Date: _____

Experiment: _____

Name: _____ Date: _____

Experiment: _____

Name: _____ Date: _____

Experiment: _____

Name: _____   Date: _____

Experiment: _____

Name: _____   Date: _____

Experiment: _____

Name: _____ Date: _____

Experiment: _____

Name: _____   Date: _____

Experiment: _____

Name: _____ Date: _____

Experiment: _____

Name: _____   Date: _____

Experiment: _____

Name: _____   Date: _____

Experiment: _____

Name: _____ Date: _____

Experiment: _____

Name: _____   Date: _____

Experiment: _____

Name: _____   Date: _____

Experiment: _____

Name: _____ Date: _____

Experiment: _____

Name: _____ Date: _____

Experiment: _____

Name: Date:

Experiment:

Name: _____ Date: _____

Experiment: _____

Name: _____   Date: _____

Experiment: _____

Name: _____ Date: _____

Experiment: _____

Name: _____ Date: _____

Experiment: _____

Name: _____   Date: _____

Experiment: _____

Name: _____   Date: _____

Experiment: _____

Name: _____   Date: _____

Experiment: _____

Name: Date:

Experiment:

Name: _____   Date: _____

Experiment: _____

Name: _____ Date: _____

Experiment: _____

Name: Date:

Experiment:

Name: _____   Date: _____

Experiment: _____

Name: _____   Date: _____

Experiment: _____

Name: _____ Date: _____

Experiment: _____

Name: _____   Date: _____

Experiment: _____

Name: _____ Date: _____

Experiment: _____

Name: _____   Date: _____

Experiment: _____

Name: _____ Date: _____

Experiment: _____

Name: _____   Date: _____

Experiment: _____

Name: _____ Date: _____

Experiment: _____

Name: _____   Date: _____

Experiment: _____

Name: _____   Date: _____

Experiment: _____

Name: _____   Date: _____

Experiment: _____

Name: _____   Date: _____

Experiment: _____

Name: _____ Date: _____

Experiment: _____

Experiment:

Name: _____ Date: _____

Experiment: _____

Name: _____   Date: _____

Experiment: _____

Name: Date:

Experiment:

Name: _____   Date: _____

Experiment: _____

Name: _____   Date: _____

Experiment: _____

Name: _____ Date: _____

Experiment: _____

Name: _____ Date: _____

Experiment: _____

Name: _____ Date: _____

Experiment: _____

Name: Date:

Experiment:

Name: _____ Date: _____

Experiment: _____

Name: _____   Date: _____

Experiment: _____

Name: _____ Date: _____

Experiment: _____